INSIDE
THE GREAT
JET FIGHTERS

INSIDE THE GREAT JET FIGHTERS

Robert F. Dorr

Motorbooks International
Publishers & Wholesalers ®

This edition first published in 1996 by
Motorbooks International Publishers
& Wholesalers, 729 Prospect Avenue,
PO Box 1, Osceola, WI 54020   USA

Previously published in Great Britain by
Windrow & Greene Ltd.
5 Gerrard Street, London W1V 7LJ

Library of Congress Cataloging-in-Publication Data Available.

ISBN 0-7603-0306-1

Printed in Hong Kong

Acknowledgements

In the course of assembling this pictorial walk around the great
jets I met a man who flew F-86 Sabres in Korea, whose son flies
F-16 Fighting Falcons in Korea. At every juncture there were
reminders that this story leaps across generations - a tale of
great planes and great men which covers the full history of the
independent US Air Force.
   The opinions expressed in this book are my own, and do not
necessarily reflect those of the United States Air Force. Any
mistakes which appear in these pages are the fault of the
author; but acknowledgement, and thanks, must be given to
many who helped. I am grateful to: Dean Abbott, Martin
J.Bambrick, Jim Beasley, George Berke, Don Brewer, David
F.Brown, Joe Buebe, Curtis M.Burns, Ed Burts, Curtis N.Carley,
Jonathan Chuck, Joe Cupido, Larry Davis, Robert Esposito,
Thomas F.Evans, Norris Graser, Robert L.Glueck, Perrin W.Gower,
Michael Grove, Michael Hagerty, Hans Halberstadt, Joseph
G.Handelman, Alex Hrapunov, Donald L.Jay, Richard Kamm, Don
Kilgus, Jon Lake, John Lauder, Franklin B.Love, Don MacGregor,
Frank Mallory, Jim McGuire, David W.Menard, Curvin Miller,
R.J.Mills, Donald R.Murphey, Dolphin D.Overton III, Richard Rash,
Bill Rogers, Bob Shane, Eric G.Smith, Richard I.Sudhoff, Warren
Thompson, and Bernard Vise.

Robert F.Dorr
Oakton, Virginia

Photography

The reproduction of the color photographs in this book varies in
balance and quality, and some transparencies show small
blemishes. We make no particular apology for this; the pictures
from the 1950s-1970s were not taken by professional
photographers during relaxed Public Affairs facilities. The
majority of them were taken by squadron personnel on active
service, with a variety of cameras and stock, under widely
varying conditions of light, temperature and humidity; and some
are more than 40 years old. That they survive at all in printable
form is in many cases remarkable.

CONTENTS

INTRODUCTION

The great jet fighters of the US Air Force are ready for inspection. We have seen them before, but perhaps never so close. On the pages which follow we examine six USAF jet fighters in detail, and in color. This celebration of the jet age will enable us to peek inside cockpits, to study bumps and bulges, and to get to know, at a fairly personal level, warplanes which define military aviation in the second half of the 20th century. Welcome to the world of the Thunderjet, the Sabre, the Super Sabre, the Phantom, the Fighting Falcon and the Strike Eagle - also known more intimately to their pilots and ground crews by unofficial names like "Hog" (F-84), "Hun" (F-100), "Viper" (F-16), and so forth. These are the jets which took us from the era of subsonic flying to the Mach 2 world of high-speed, high-G aerial combat; from the lean years before the Korean War, to the lean years after the Gulf War....

One way to think of the time span covered by this pictorial glimpse at the great jet fighters is to think of the world events which measured off our lives while these aircraft were taking to the air for the first time.

On 28 February 1946, Wallace Lien took off from Muroc Army Air Field, California, in the first Republic XP-84. The following year, on 18 September 1947, the US Air Force became an independent service. The year after, on 11 July 1948, the designation "P" for pursuit became "F" for fighter. The following year the death of a test pilot called Edwards gave the airfield at Muroc a new name. The Republic F-84 Thunderjet became an air-to-ground "flying dump truck" in the Korean conflict of 1950-53, during which it carried just about every kind of bomb and rocket in the USAF arsenal, and flew tens of thousands of combat missions against the North Korean and Chinese forces. Thunderjets even managed to shoot down a handful of MiG-15s. The straight-wing F-84 became a bulwark of NATO and other Allied air forces, and served on faithfully in the ranks of the Air National Guard.

On 1 October 1947, World War II ace and civilian test pilot George Welch took off from Muroc in the first North American XP-86. No one will ever know, and nobody can ever prove it, but many North American engineers believe that they heard sonic booms during early test flights, and think that Welch probably flew faster than sound before test pilot Captain Charles E. Yeager made the first supersonic flight in the Bell XS-1 on 14 October 1947. The North American F-86 Sabre became the air-to-air ace of the

Korean War; starting out with some initial disadvantages vis-a-vis the brilliant MiG-15, the Sabre decisively defeated the Russian jet and ran up a ten-to-one kill ratio in massive air battles along the Yalu River. More than 9,000 examples of all models of the Sabre were built; it too served in the air forces of America's allies, and in the Air National Guard.

On 25 May 1953, George Welch made the first flight at what was now Edwards Air Force Base of the North American YF-100A. This would become the world's first fighter capable of supersonic speed in level flight on a sustained basis. The North American F-100 Super Sabre was essentially a creature of the Cold War; it made its mark along the frontiers of the free world in Europe and Asia, confronting the Soviet Union and China at every turn. For years, long before the USAF reverted to camouflage-painted livery, the Super Sabres gallivanted about the upper atmosphere in their gleaming natural metal finish. Only late in its service life, long after it had earned the affection of a whole generation of pilots, did the F-100 go to war in Vietnam. The "Hun" was exported only in small numbers, although it was flown by the Air National Guard for many years; and those who cherish fond memories of it believe that it never quite received the recognition it deserved.

On 27 May 1958, civilian test pilot Robert Little made the first flight at Lambert-St.Louis Municipal Airport, Missouri, of the prototype McDonnell F4H-1 - the only aircraft in this half-

dozen to have been developed by the US Navy. Known at first as the F-110 until the system of designating US military aircraft was overhauled on 1 October 1962, the McDonnell F-4 Phantom became the standard against which every other warplane of its generation had to be measured. With two men, two engines, and - at first - no guns, the Phantom was not pretty (in fact some called it "Double Ugly"); but the Phantom defined the central years of the Cold War and the conflict in Vietnam to such an extent that its appearance, perfectly combining form and function, would come to inspire great affection. Its manufacturer became McDonnell Douglas in 1966 (although the McDonnell Aircraft Company retained its separate identity within the parent firm throughout the lifetime of the F-4). The Phantom served on land and by sea, in the colors of a dozen nations; in time it passed into the armory of Guard and Reserve units; and in its RF-4C reconnaissance and F-4G "Wild Weasel" guises it flew many combat missions during the Gulf War of 1990-91 - more than 30 years after its maiden flight. To the US Air Force the Phantom was unique; there was never again to be anything quite like it.

On 20 January 1974, test pilot Phil Oestricher made an inadvertent first flight at Edwards in the General Dynamics F-16, followed by an intentional maiden flight on 2 February. The manufacturer became Lockheed Fort Worth in 1992, and Lockheed Martin in 1995. The Lockheed Martin F-16 Fighting Falcon (alias "Viper" or "Lawn Dart") is widely viewed today as the world standard for a multi-role fighter; it serves in no fewer than 16 friendly air forces, and five of those countries manufacture or assemble the aircraft themselves. The F-16 introduced a plethora of high-tech inventions, including its computer-guided "fly by wire" system, a sidestick controller in place of a conventional centrally mounted control column, and a slightly reclined ejection seat. US Air Force F-16s served with great success in Operation Desert Storm; and are today in the vanguard of USAF, Air National Guard and Air Force Reserve fighter operations.

On 11 December 1986, test pilot Larry Walker was at the controls for the first flight of the first F-15E Strike Eagle, developed from a McDonnell private venture which, in turn, was based on earlier Eagle variants. The McDonnell Douglas F-15E Strike Eagle has replaced the General Dynamics F-111 Aardvark as the USAF's deep strike, air-to-ground warrior; it equips fighter wings based

in North Carolina and England, and composite wings in Idaho and Alaska. The most recent jet fighter to attain operational status with the USAF, it is the basis for Eagle variants being manufactured for Israel and Saudi Arabia, although the 109th and last F-15E ordered for American service has now been delivered.

*   *   *

1946, 1947, 1953, 1958, 1974, 1986. . . The changes witnessed by these six jet fighters involve far more than simply speed. The first F-84 Thunderjet did not have an ejection seat, while the ACES II ejection seat in the F-16 and F-15E provides almost certain assurance of survival in an emergency, even at zero speed and zero altitude. The first XP-86 was rumoured to have gone supersonic; the F-16 spearheaded the first generation of warplanes able to *accelerate while flying straight upwards.*

The F-84 and F-86 were flown in the world's first jet-vs.-jet air war by men who had become prop-fighter aces in World War II, half a century ago; today the F-15E Eagle's pilots include young women who weren't yet born when the Beatles broke up. In today's F-16 and F-15E we have aircraft which for the first time are structurally able to handle more Gs - to absorb greater stress during abrupt manoeuvering - than the bodies of the humans who fly them.

Greatness invites mimicry; inevitably, the great warplanes on these pages have close relatives not included in our pictorial tribute. The straight-wing F-84 Thunderjet was developed into the swept-wing F-84F Thunderstreak and RF-84F Thunderflash series, separate aircraft types which are outside the scope of this examination. The F-86 Sabre inspired the FJ Fury which flew with the US Navy and US Marine Corps, as well as Canadian and Australian offspring. The F-4 Phantom began with the US Navy, and served Navy and Marine squadrons as admirably as it did the Air Force. The F-15E came along only after the success of the earlier, air-to-air fighter F-15 Eagle variants.

When we set out to create this book, we decided from the start that we would depict only *operational* warplanes in actual squadron service. In many cases the aircraft shown on these pages are not merely operational: they are in combat. There are no experimental prototypes here, no replicas or museum pieces or gate guards. Every one of these jets is a real warplane, doing a real job.

CHAPTER 1

Republic F-84 Thunderjet

(Left top) "Paige Ann", Republic F-84E-15-RE Thunderjet (49-2407), was serving with the 49th Fighter-Bomber Wing when she was photographed at Suwon, South Korea, in 1951. Soon after the first flight of a prototype XP-84 on 22 February 1946 the Thunderjet showed its stuff by establishing a new US national speed record of 611mph (983kmh). The F-84D was the first version to see combat in Korea, soon followed by this F-84E, which had the fuselage lengthened by 12 inches, and subsequently by the F-84G. *(Robert L.Glueck)*

(Left bottom) An F-84E of the 31st FB Wing at Turner Air Force Base, Georgia, in 1951. Commanded by air ace Col.David Schilling, the 31st pioneered long range jet deployments using air-to-air refuelling. This Thunderjet has been fitted with two RATO (rocket assisted take-off) units to boost its take-off run; solid propellant units made by Aerojet, these weigh 170lbs (77kg) each, produce 1,000lbs (454kg) of thrust, and burn for ten seconds. The RATO units are fitted into electrically controlled hooks under the rear fuselage; they are jettisoned after take-off, and are not re-used - all the pilot has to do is hit his jettison switch. The F-84E carried 230-USgal. wingtip tanks, and could also carry two of these on the underwing pylons; these tanks were manufactured by Republic, and no other fighter used them. *(Dolph Overton)*

(Right) The well-dressed Thunderjet pilot of the 1951 season: 1st Lt.Bill Rippy of the 429th FB Sqn., 474th FB Wing at Kunsan air base, South Korea. He wears helmet, parachute, "Mae West" life preserver, G-pants, khaki flying coveralls, leather gloves and heavy boots. Note, behind him, the dive brake of his F-84G with its four rectangular holes; some early F-84Gs may have had the earlier "Swiss cheese" type. *(via Warren Thompson)*

(Left) Not often seen from this angle, a Republic F-84E of the 111th FB Sqn., 136th FB Wing is scrutinized by maintenance men at K-2 Taegu air base in late 1952. The fuselage tapers cleanly into a small, almost circular tailpipe for the 5,000lb (2,268kg) thrust Allison J35-A-21A axial flow turbojet engine. The underside of the wingroots is rather cluttered; seen here bracketing the red "no step" sections on the wing trailing edges are the inboard pylons capable of holding 1,000lb (454kg) M-117 bombs; and the connector doors which close over the main landing gears when they retract. Note the dihedral on the straight wing of the F-84, a feature not found on later jet fighters. (Franklin B.Love)

(Below left) This F-84E Thunderjet (49-2361), nicknamed "Donna" and wearing "buzz number" FS-361, was hit by anti-aircraft shells forward of the pilot's windshield during a mission flown by Capt. Robert F. Crutchlow of the 111th FB Sqn., 136th FB Wing. One shell entered the nose of the aircraft and exploded in the pilot's face; temporarily blinded, Crutchlow still managed to bring his plane back to Taegu with the help of two brother pilots flying off his wing. This perspective of the nose shows that even at the best of times the pilot had a narrow field of vision ahead and limited visibility all around, thanks to the narrow windshield and braced canopy. Note the anti-glare panel forward of the cockpit; wear and tear on the paintwork around the F-84E's perfectly circular air intake; and the gun ports for the pair of .50cal. M3 Brownings each side of the upper nose. (Franklin B.Love)

(Below) Laden with bombs, Republic F-84G Thunderjets head north to inflict harm on North Korea. The lead aircraft is being flown by 1st Lt. Jim "Suitcase" Simpson of the 69th FB Sqn., 58th FB Wing, out of Taegu. The F-84G was the first fighter with a fuel receptacle for the "flying boom" employed by some KB-29 Superfortress tanker aircraft - although unfortunately none were available in Korea. The box-finned 2,000lb (907kg) high explosive bombs are typical of World War II stock which remained available for use in the Korean War, and for years afterwards. The F-84G is often called an "interim" aircraft anticipating the swept-wing F-84F Thunderstreak, which was long delayed in development. In fact no less than 3,025 of the G-model were manufactured (789 of them for the USAF), making it the most widely built version of the F-84.

(Above) In January 1952 Thunderjets stationed at Misawa air base in northern Japan flew the first of a series of "skat" or shuttle missions - striking a target in North Korea, landing at Taegu in the south, then launching a second mission from which they returned direct to Misawa. During the stop-over at Taegu Korean workers assist US armorers in loading 1,000lb HE bombs onto Thunderjets of the 159th FB Sqn., 116th FB Wing - an Air National Guard unit mobilized for the fighting in Korea.
(Martin J.Bambrick)

(Right) A glimpse of retro-fitted braced canopy detail is offered by this shot of "Mrs.Murphey", a 58th FB Wing F-84E at Taegu in 1953. The nickname on the right side of an aircraft is usually applied by the enlisted crew chief or armorer, often with something less than the most exquisite artistic skill. The fit of the canopy in its rails was much smoother than it looks here, though not always airtight or free of leaks, which could annoy a pilot flying at altitude.
(Donald R.Murphey)

(Above) "Miss Patty Ann/ Leading Lady" is a Republic F-84G-26-RE Thunderjet (51-16657A), flown by Lt.Fritz Frazier of the 69th FB Sqn., 58th FB Wing from Taegu in early 1953. The F-84G had from the outset the braced or "stripped" canopy which was later retro-fitted to some F-84E models. Early in the production run F-84Gs were built with the perforated "Swiss cheese" under-fuselage dive brakes shown here; later G-models had brakes with four rectangular holes. The large inboard pylon carried a 230-USgal. droptank, or a bomb. What appears here to be a smaller pylon slightly outboard is actually a partial cover for the main landing gear wheel, with a taxi light in the extended position. Beneath the prominent buzz number FS-657 is a squareish cooling door for the engine. The F-84G mounted six Browning M3 .50cal. (12.7mm) machine guns - four in the nose and one in each wingroot; the port for the wingroot gun is visible here above the pylon. (Donald R. Murphey)

(Above) Line-up at Tsuiki, Japan, where major maintenance work was performed on aircraft fighting in Korea. Partly visible in the background are a Curtis C-46 Commando and a North American F-51D Mustang, both typical of an era when more warplanes had propellers than jet engines. The two Republic F-84E Thunderjets are relatively clean considering the primitive conditions in which work had to be carried out. The F-84E model made its first flight on 18 May 1949 and was available in considerable numbers by the time the Korean fighting began. The last of 843 F-84Es rolled off Republic's Farmingdale, Long Island, production line in July 1951. (Robert L Glueck)

(Above) An example of rather gaudy crew chief's or armorer's "nose art" from the right side of an F-84G Thunderjet in Korea, where the name or artwork chosen by the pilot appeared on the left side only. (via Robert F.Dorr)

(Below) F-84G of the 311th FB Sqn., 58th FB Wing taxying out at Taegu for a combat mission in 1953. The position of the nose art this far forward, ahead of the buzz number, is very unusual. "Buzz numbers" were intended to allow quick identification of individual aircraft (the term comes from the notion of angry citizens telephoning the Air Force to complain about being "buzzed" by low-flying jets). In the busy 1950s a letter suffix, like the A following FS-106 here, was sometimes necessary because more than one airframe of the same type would have the same "last three".

(Left) A signboard at Taegu air base tells the world that two F-84G fighter-bomber wings (the 49th and 58th, indicated by the anachronistic buzz number in the artwork), are in residence. The design of the straight-wing F-84 was deliberately conservative, and the Air Force bought it as insurance against the possible failure of the more radical swept-wing F-86 then under development. Thunderjet pilots tended to speak of their flying school classmates now occupying Sabre cockpits with a degree of disdain; one complained that his biggest fear was of being struck by wingtanks dropped by Sabres "flying up to the Yalu River to tool around the sky and do a whole lot of nothing, while we were out winning the war..."
(John Lauder)

(Below left) "Barbara M." was (and still is) the bride of 1st Lt. George Berke, who flew F-84E and F-84G Thunderjets with the 69th FB Sqn., 58th FB Wing out of Taegu. Note the blue trim on the nosewheel door (the squadron color), and the old-style dive brake. A pilot usually had to wait some weeks after joining a squadron before he got his name on the canopy rail of an assigned aircraft.
(George Berke)

(Below) This close-up of F-84G "The Vicious Virgin", FS-169A, shows a good view of the pilot's flying gear, and the braced canopy. In the shadows can be seen a pair of bombs; and the JATO (jet assisted take-off) unit used to give the Thunderjets a boost during take-off with a heavy load in the heat of the Korean summer, for which task the F-84 was rather underpowered - two bottles were slung externally from the lower rear fuselage in a bracket with a jettison release. Yellow squares on the fuselage above the wing indicate access panels.
(via Robert F.Dorr)

In any conflict aircraft are lost in non-combat mishaps - usually, about as many as are lost in action. It is not known why this F-84G (51-10309) of the 49th FB Wing burned on the taxiway at Taegu, or whether the pilot was inside when it happened. Judging from the photos, the fire appears to have started in a droptank on the right inboard pylon. The pyrotechnics are somewhat unusual: notwithstanding Hollywood dramas, aircraft do not automatically explode when they catch fire, and they do not always burn with such searing intensity that large amounts of metal are literally consumed. The overall safety record for the F-84 Thunderjet was better than that of several other jets of the Korean era, including the F-86 Sabre, but not as good as that of the Lockheed F-80 Shooting Star.
(via Robert F.Dorr)

(Above) A neat comparison of two straight-wing jets flown by the US Air Force in Korea. The Lockheed T-33A Shooting Star trainer, left, was developed from the F-80 fighter, and was similar in size and general flying characteristics to the F-84G Thunderjet. Most fighter squadrons in Korea eventually acquired a T-33A for area familiarization flights, VIP travel, and general "hack" duties. The Thunderjet at rest on the pierced steel planking in this view seems to be bombed up and ready to go on a mission.
(via Robert F.Dorr)

(Below) The standard cargo of the heavy-hauling F-84G: 2,000lb HE bombs left over from World War II are hauled around Taegu air base on primitive trailers in 1952. Much of the task of loading this ordnance involved winches, chains, and human muscle. Note the Thunderjet in the background, which has been "broken". The tail section minus the engine is facing directly towards the camera; the forward section faces left three-quarter towards us. Because Korea had been occupied by Imperial Japan from 1905 to 1945 it had a well-established infrastructure of roads, railways, and airfields; however, the Japanese had been defeated before they had a chance to build modern facilities on the latter, and consequently virtually all Allied aircraft maintenance in 1950-53 had to be carried out in the open - in extreme heat during summer and in near-Arctic cold during winter. It was a tough time and place to be a crew chief or an armorer.
(via Robert F.Dorr)

(Right) Close-up of 1,000lb HE bomb hanging from the generously sized inboard pylon of an F-84E. It is attached by four principal lugs points and four more wiring attachment points. Note the bomb's modified box-shaped fin, and the fuse spinner (wired here to render it safe); the large main gear taxi light; and the bulge beneath the tail which would absorb any impact if the pilot "over-rotated". F-84s dropped over 55,000 US tons of explosives during the two and a half years that they were operational in Korea. In one memorable mission a Thunderjet pilot from the 49th FB Wing successfully lobbed two bombs into the mouth of a railway tunnel to blow up a train entering it from the other end.
(via Robert F.Dorr)

(Left) Staff sergeant of the 8th FB Sqn. "Black Sheep",part of the 49th FB Wing at Taegu, demonstrating the removal of access panels in the left center fuselage of F-84G (51-10460). Being able to reach inside the aircraft at this location facilitated work on electrical wiring associated with the engine, which could otherwise be serviced only when removed completely. (via Robert F.Dorr)

(Opposite top) Seen from the front, "cleaned up", the F-84 has especially pleasing lines - a masterpiece of functional design. Note the divider which bifurcates the flow of air through the intake, around both sides of the cockpit and back to the engine. The narrow windshield did not afford the best visibility; but the straight-wing F-84s had an unusually wide track undercarriage, which gave superb taxying characteristics.
(via Robert F.Dorr)

(Above) Here's what it looks like when it's broke... At Taegu technicians from the 58th FB Wing are bringing up to speed the front end of an F-84E of which the rear section, including the 5,000lb thrust Allison J35-A-21A turbojet, has been removed for serious maintenance. The man sitting ahead of the cockpit is working on the nose guns and electrical wiring, courtesy of the long, narrow "hood" which could be raised to give easy access.
(via Robert F.Dorr)

(Above left & right) F-84G Thunderjet of the 8th FB Sqn. taxying out on a Korean airfield with the red dust of summer kicking up through the PSP track; and a blurred but atmospheric snapshot of a pair of G-models lifting off. While details are not identifiable from this photo, it does confirm that "nose art" was commonly painted on the fuselage behind the buzz number. *(via Robert F.Dorr)*

(Left) Thunderjet (51-16687) of the 429th FB Sqn., 474th FB Wing at Kunsan air base, Korea, shortly after the armistice brought an end to hostilities in July 1953. Note the small blue-painted training bomb; the early type dive brake, with its 56 round, two-inch perforations, partly visible behind the low-technology metal boarding ladder; and the auxiliary power unit attached by an umbilical to the right side of the aircraft for engine starting. *(Don Brewer)*

(Left & below) Line-up of Thunderjets at Taegu base during the Korean War; and a portrait of F-84G (51-10414), which from its three-color stripes is almost certainly a wing commander's mount - but which wing? The markings of Thunderjet units in Korea are a subject of considerable confusion, in part because group, wing and squadron designations could change overnight with no corresponding change in the paint-job applied to the aircraft. The author has been unable to identify the insignia on the tail of this F-84G. *(via Robert F.Dorr)*

(Above) Hard work on an F-84G which is temporarily missing its canopy; given the participation of the pilot in whatever is going on here, still wearing his parachute harness, it is possible that this Thunderjet lost its canopy in combat and returned to base without it. Note the standard color of the anti-glare panel painted on the nose and − unusually − right back along the spine to protect the pilot's eyes from sun dazzle off the polished bare metal; the F-84 was one of the few jets to carry the panel on the rear fuselage. *(via Robert F.Dorr)*

CHAPTER 2

North American F-86 Sabre

(Above) Over Korea, 1954: 1st Lt John Wilson flies North American F-86F-30-NA Sabre (52-4499) belonging to the 8th Fighter-Bomber Wing's 80th Fighter-Bomber Squadron, the "Headhunters" (identified by the yellow tail trim). A Sabre from this squadron scored the last MiG-15 kill of the Korean War era in an incident in 1955. Wilson's Sabre has the wing fence just above the leading edge which is an identifying feature of late F-86F models with the "6-3 hard" wing, dispensing with the leading edge slats found on F-86A, F-86E and early F-86F Sabres. The term comes from the fact that the leading edge was extended 6 inches at the root and 3 inches at the tip; this made the Sabre more difficult to handle in the airfield pattern, but the most manoeuverable dogfighter in the world at that date, superior to the MiG-15 in every respect.

The broad yellow fuselage and wing bands edged black were to distinguish the F-86 at a distance from the MiG-15 of superficially similar plan. The red line around the fuselage bisecting the national insignia marks the compressor of the 5,800lb (2,630kg) thrust General Electric J47-GE-27 engine. The two black lines around the lower rear fuselage, just visible under the buzz number, mark the attachment point for the dolly used when the Sabre's aft section is separated from the rest of the fuselage. *(John Wilson)*

(Below) The black and white identification bands on this Sabre were used by the 4th Fighter-Interceptor Wing only during the very first months of Sabre operations in Korea. The early windshield shape is another clue that (49-1140) is an early F-86A, the model which first entered combat against the MiG-15 in December 1950; and the gray/white laminated lip around the air intake was also a feature of early Sabres. This aircraft was flown by Major Robert Moore and other Korean War air aces before failing to return from a mission along "MiG Alley".
(via Robert F.Dorr)

(Above) Photographed on 28 February 1950 - four months before the surprise Communist invasion triggered the Korean War - US Air Force dependents Jim Prebbanow and Jean Abay pose in front of a factory-fresh North American F-86A Sabre (48-159) at Williams AFB, Arizona. Behind Jim, in a blaze of blue, is the distinctive "Hat in the Ring" insignia of the 94th Fighter Squadron, which claims among its alumni Capt.Eddie Rickenbacker, the 26-kill top US air ace of World War I. The squadron was part of the 1st Fighter Group at March Field, California, and was visiting Williams that day. Early F-86As had flush rectangular covers over the six gun ports in the nose, designed to reduce drag; this proved a troublesome feature and was quickly discarded. Behind Jim's elbow the access door for the ammunition supply dangles open; at the rear fuselage a dive brake hangs open, overlapped by the national insignia. Early F-86A Sabres were prone to various glitches, but most of these were resolved by the

time the first examples arrived in Korea in December 1950.
(Joe Buebe)

(Top right) The classic view of a Sabre, familiar to the ground personnel who gathered near the end of the runway to watch a mission take off. This F-86A of the 335th FI Sqn., heading north to look for trouble along the Yalu, displays the yellow fuselage and wing bands which became standard for Sabres, plus the fin band which specifically denoted the 4th FI Wing. The "Chiefs" squadron emblem is seen on the left forward fuselage; the droptanks are standard; note also that the nosewheel has retracted but its big door has yet to traverse into the closed position.The F-86 was originally conceived as a straight-wing aircraft, and a full scale mock-up in that configuration was completed by North American in 1945. When wartime German data on swept wings became available, the then-US Army Air Forces made the courageous decision to go ahead with the less proven

but more advanced wing shape. Although it required extensive development, the resulting aircraft was a world-beater.
(Martin J.Bambrick)

(Right) "Phyllis" - named by a ground crewman - is an F-86E Sabre (50-594) of the 51st FI Wing photographed at Suwon late in 1951; pilot Phil Norton left "his" left side of the aircraft in unsullied natural metal. Here armorers are working on the starboard battery of three M3 machine guns; somewhat prone to jamming, they were relatively easy for the armorers to load, unload, or even remove - though they had to do so with a minimum of support equipment.
(Phillip A.Norton)

(Above) Capt.Karl K.Dittmer of C Flight, 335th FI Sqn. ("Chiefs"), 4th FI Wing, poses in front of his own artwork on "Betty Boots" at Kimpo in 1952; note the kill star for a MiG-15. His right knee is up against the ammunition pans for his F-86E Sabre's six .50cal. M3 Brownings, where an access panel has been removed; the guns were armed with an impressive total of 1,261 rounds. The pilot's right hand is on one of two collapsible steps used to facilitate climbing up to the cockpit.
(Karl Dittmer)

(Top right) Karl Dittmer was handy with a paintbrush as well as an F-86; he embellished five Sabres of his unit with nose art, and is seen here at Kimpo in 1952 posing beside one of them: F-86A (49-1272), flown by his wingman 1st Lt.Martin Bambrick - who also got himself a MiG. This view of "Wham Bam" illustrates the different tones of the various panels in the forward fuselage. Note also the khaki flight suit and silver-trimmed Air Force blue cap with pin-on rank bars.
(Karl Dittmer)

(Right) F-86E Sabres (foreground, 50-625) of the 4th FI Wing at K-14 Kimpo air base outside the South Korean capital of Seoul in 1951 or 1952; the "Chiefs" emblem identifies the 335th FI Squadron. The large scoreboard of red stars low on the right forward fuselage presumably does not represent individual victories credited to pilots of this particular Sabre, but the cumulative kill score of the unit.
(Karl Dittmer)

(Above) K-14 Kimpo, late 1952: the red bird in a black-bordered yellow circle looks like a rooster, but was in fact the emblem of the 334th FI Sqn., 4th FI Wing, who were known as the "Pissed-Off Pigeons". This Sabre (52-2875), one of the 60 Canadair F-86Es built for the USAF north of the border, displays to advantage the leading edge slats found on all Sabre models between the F-86A and early F-86F. Extending these slats increased the wing area and improved performance at low speed, as when in the airfield landing pattern. Note also the APU used to start up the Sabre's J47 turbojet engine.

The 4th FI Wing always had at least one British, one Canadian and one US Marine exchange pilot on the strength. This particular Sabre was assigned to Squadron Leader Eric G.Smith, RCAF, a veteran of World War II combat in De Havilland Mosquitos who flew a tour with the 334th FI Sqn.; he flew wing on Capt.Leonard W.(Bill) Lilley, a USAF ace who shot down seven MiGs.
(Eric G.Smith)

(Top right) A pair of early F-86F Sabres (51-12963 & 51-12594) on the prowl over Korea; among the first F-models to arrive in theater, they are flown by 4th FI Wing pilots Capts.Clifford D.Jolley (later an air ace) and Karl K.Dittmer. While not apparent to the naked eye, these Sabres are equipped with RATO bottles in the lower fuselage, intended to provide a burst of speed to improve closure rates with the MiG-15. This experiment produced mixed results, but apparently contributed to at least one MiG kill. Though the Sabre was employed as a fighter-bomber late in the war, it took a considerable time to prove itself in that role; the great majority of its missions were flown in this configuration, during the fight to win and keep air superiority.
(Karl Dittmer)

(Right) F-86As and F-86Es of the 334th FI Sqn. undergoing maintenance at Kimpo, 1952; note in background an aircraft with its rear fuselage removed for engine access. This view was taken by USMC exchange officer Capt. Richard Rash, who flew with the sister 336th FI Sqn., and learned patience from the experience ("How would you like to be thrown in with a group of Air Force jocks if you were wearing a name tag saying 'Dick Rash'?. . .) Rash was also appalled - unsurprisingly - by the minimal facilities available to the main-tenance people at K-14.
(Richard Rash)

(Below) 1st Lt.Don MacGregor of the 25th Fl Sqn., 51st Fl Wing, photographed in the cockpit of an F-86E Sabre at K-13 Suwon air base in mid-1953. The Sabre was one of the last fighters in which the parachute was attached only to the pilot rather than to the ejection seat. The "bone dome" worn in the 1950s was miserably heavy and often ill-fitting by today's standards; the oxygen mask could chafe painfully at high altitude. Despite the dark tones of the photograph this angle does give some idea of the clutter created by the F-86E's A-4 ranging gunsight. *(Don MacGregor)*

(Right) The cockpit of the F-86F Sabre was cramped, cluttered, but functional. This photo shows the place of work of 1st Lt.William Osborne of the 39th Fl Sqn."Cobras", 51st Fl Wing, based in 1953 at Suwon. Osborne has hung his helmet over the rim of the canopy bow with the tube for the oxygen mask dangling in front of the instruments. Note the leather gloves - these were de rigeur for pilots in the days before Nomex fire-retardant fabric. The firing button for the Sabre's six .50cal. M3 Browning machine guns is on the handgrip of the control stick. Note just visible at the right the five red-painted stars under the canopy rail which mark this Sabre as an ace MiG-killer. *(via Warren Thompson)*

(Bottom right) Kimpo, 1952: Dick Warren - a USMC officer on exchange with the 4th Fl Wing, and identified as a Marine by the "cover" on his head and the Navy-style "Mae West" - poses as if performing a pre-flight inspection of a late F-86E or early F-86F. From this angle the somewhat narrow track of the Sabre's main undercarriage is apparent; and note the A-3 starter cart. The red line painted on the PSP is to help the ground crews to park the Sabres evenly; having taken command of the air long ago, the 4th have no fear of being strafed on the ground, and make no use of protective bunkering. *(Karl Dittmer)*

(Right) Another peek into the cockpit, this time of an F-86E of the 39th FI Sqn.("Cobras"), 51st FI Wing at Suwon in 1953. The red-tagged warning labels used to cage the gunsight are little different from those in use today. The console between the pilot's knees is especially cluttered, and the analog flight instruments in their shielded dials required the pilot to look down at a considerable angle to get some readings - characteristics which have been carefully designed out of today's fighters, with their digital instruments and "head-up displays". This Sabre - see extreme right - racked up three MiG kills. The five-pointed, two-and-a-half-inch, insignia red star signifying an aerial victory was technically inaccurate, being the Soviet insignia rather than that of the Chinese or North Korean air forces who were (officially) our adversaries in 1950-53; but it was fairly standard on US fighters nonetheless - in those innocent days it was broadly felt that a Commie was a Commie. . . The instinct proved correct: we now know that large numbers of Russian MiG pilots fought in Korea in false uniforms and markings.
(Dean Abbott)

(Left) Seen in a sandbagged dispersal pen at K-13 Suwon, with external fuel tanks and M-117 bombs in place, "Terrible Turtle" is an early F-86F Sabre (52-4446) with leading edge slats. It was flown in 1953 by 1st Lt. Perrin W.Gower of the 35th Fighter-Bomber Sqn., 8th FB Wing - the first unit to employ the Sabre in the fighter-bomber role. (Perrin W.Gower)

(Top left) Surrounded by pools of muck reflecting the primitive conditions of the time, 2nd Lt.Bernard Vise taxies in at K-13 Suwon following a fighter sweep along MiG Alley in his F-86E (50-598),"My Best Bett". The blue fin stripe above the black checkers identifies the 16th Fl Sqn., 51st Fl Wing; red identified the wing's 25th Fl Sqn., yellow the 39th. Vise remembered that there was a severe wingtank shortage during his combat tour in Korea, although he has them here. The F-86E introduced the "all-flying tail" - the horizontal stabilizer which moved as a single slab. The 51st Fl Wing began converting from the F-80 to the F-86E

Yalu. One pilot noted that at $1,900 a pair of these tanks cost exactly the same as a new 1951 Chevrolet coupe - "I felt I was throwing away the automobile I wanted every time we sent those gleaming metal tanks tumbling towards the ground".
(Karl Dittmer)

(Left & above) The "Newark Fireball" was an F-86E Sabre (51-2894) flown by Capt.Henry A.Crescibene of the 335th Fl Sqn. "Chiefs", 4th Fl Wing out of Kimpo. Hank Crescibene got one MiG-15 kill before leaving the US Air Force to fly F-84F Thunderstreaks and F-105 Thunderchiefs for Republic. Aircraft in the late F-86E series were actually F-86Fs in every respect except for the engine, and were so redesignated after retro-fitting of the F-model's more powerful version of the J47 turbojet. The standard 275-gallon droptanks were carried by all Sabre models, and were often jettisoned prior to fighter sweeps along the

in November 1951, becoming the second Sabre wing in Korea. Contrary to widely published accounts (and the inaccurate recollections of some of its members), the 51st never flew the F-86F model.
(Bernard Vise)

(Left, bottom left & below)
"Wyoming Thunder" is an F-86E Sabre (51-2734) of the 25th Fl Sqn., 51st Fl Wing which had Airman First Class Donald C.Porter as crew chief, and a succession of pilots. When these portraits were snapped "Thunder" was the mount of Major E.W."Hap" Harris; and every pilot, crew chief and armorer had his name stencilled on a red field below the left side of the cockpit. Note details of feed arrangement for the three .50cal. machine guns exposed by the absent access panel.
(Donald C.Porter)

(**Below**) An example of the rarest version of the Sabre built, this RF-86A Sabre (48-217) is one of the Project Ashtray aircraft converted in Japan for photo-reconnaissance missions. It belongs to Col. Edward S. Chickering's 45th Reconnaissance Sqn. "Polka Dots", part of the 67th Reconnaissance Wing at Kimpo air base. Nicknames on the right side were usually applied separately by crew chiefs and armorers, hence the two different names "Every Man a Tiger" and "Honeybucket". There is some dispute about the exact number of RF-86As converted, but there were no more than about half a dozen. Less rare, though far from common, was the RF-86F model, which differed little in appearance.
(*via Robert Esposito*)

(**Right**) Although the red tail stripe on F-86E (51-2832) at Suwon identifies the 51st FI Wing's 25th FI Sqn., the tricolor nose stripe marks "Karen's Kart" as the mount of fighter wing commander Lt. Col. Al Kelly, crewed by A/3C R.E. Wright, both of whose names appear on the nose. In 1952 the USAF replaced Army-style enlisted ranks with its own terminology - "Airman First Class", etc. Although the broad yellow fuselage band became a standard identification aid carried by nearly all Sabres, it was devised by the 51st FI Wing. (*Curvin Miller*)

(**Bottom right**) A new canopy is installed on an F-86E at Suwon in 1953. (*Bill Rogers*)

(Above) F-86Es of the 25th FI Sqn. lined up on the matting at Suwon in 1953; despite the rigors of war, the crew chiefs and maintenance personnel of the 51st Wing kept their Sabres gleaming, the natural metal finish often polished to a higher sheen than the day they left the factory. The foreground aircraft (50-649) has its USAF serial number incorrectly displayed, according to the conventions of the time: as the 649th aircraft ordered by the service in fiscal year 1950 (1 July 1949-30 June 1950) this Sabre should have a tail marking beginning with only the final digit of the year - "0649". The droptanks shown here began to appear late in the war, and are described by one source as "F-80 style Misawa tanks". While they may well have originated at Misawa, the northern Japanese base where improved tanks were developed, they showed up in later years in places as far afield as Saudi Arabia. *(Curvin Miller)*

(Left) Led by the 25th FI Sqn. commander (note double red nose stripes), F-86Es taxi out for a mission from Suwon in 1953; note non-standard Japanese-made droptanks, and the black laminated cover for the range-computing gunsight at the top of the nose intake. In the final year of the Korean War Sabres fought in some of the biggest air battles since World War II; these high-speed clashes, often bewilderingly fleeting for the pilots of this first generation of machine gun-armed jets, could see many dozens of Sabres and MiGs duelling each other in the narrow airspace just south of the Yalu.
(Curvin Miller)

(Above) K-14 Kimpo air base: the year is 1954, and pilots of C-119 transports like that just visible at left were spending most of their time supporting the besieged French garrison of Dien Bien Phu in North Vietnam, in the face of flak as intense as any the USAF had encountered in the Korean War which had ended in armistice a few months before. "Julie" is an F-86F Sabre (52-4422) of the 8th Fighter-Bomber Wing, photographed by C-119 pilot Ed Burts - who had flown fighters in World War II, but was seeing a Sabre here for the first time. His photo shows the engine start-up procedure, with the pilot and crew chief communicating via hand signals. Evident here are various details including the upper part of the nose landing gear; problems with this assembly were never fully solved, and it often collapsed during a rough landing. The three red stripes identify the aircraft flown by the commander of the 8th Fighter-Bomber Wing's 36th FB Sqn., "Panthers", which converted from the Lockheed F-80C Shooting Star in early 1953 and flew the F-86F in the closing months of the Korean fighting.
(Ed Burts)

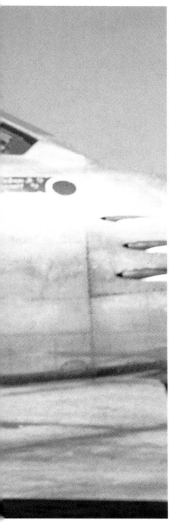

(Opposite top) An oddity in a geographically straightforward war, the island of Paengyongdo ("White Flower Island", better known to Americans as "PY-do") lay north of the 38th Parallel; its hard-packed Yellow Sea beach sand offered an emergency landing strip for battle-damaged Sabres. In this photograph, taken at some date between January and May 1953, two F-86Fs of the 336th FI Sqn. "Rocketeers" (51-12976 & 51-2863) are preparing to take off from the beach after having taken temporary refuge, watched by Marines of the West Coast Islands Defense Unit (WCIDU); not visible here is the Sikorsky H-19 helicopter which hovered in the background, alert for trouble, during such take-offs. The island remained in friendly hands after the armistice of 27 July 1953, and served as a handy intelligence-gathering outpost.
(Richard I.Sudhoff)

(Left) Major Ed Fletcher (center), commander of the 71st FI Sqn. at Suffolk County, New York, appears here with several of his pilots in front of an early F-86A Sabre. Although most of the men in this photo are Korean War veterans and the shot may have been taken shortly after the war in 1954, it is no secret that Air Defense Command squadrons enjoyed priority in receiving Sabres from the factory, even ahead of the units which were fighting and dying in Korea. Since the Sabre was one of the few fighters in history designed for a single mission - daytime, good-weather air superiority - it is difficult to imagine how much damage ADC might have hoped to inflict on attacking Soviet bombers with a fleet of machine gun-armed F-86As. The squadron did eventually convert to the radar-equipped all-weather F-86D interceptor.
(Martin J.Bambrick)

(Above) Unfortunately, the author has been unable to locate any record of who preserved this shining sunset moment on the 4th FI Wing's base at Kimpo. The photographer was presumably the pilot whose helmet rests on the canopy bow; but his name, his squadron, and his story are four decades behind us now.

Interestingly, Kodak color slide film of the kind on which this image is preserved has been readily available in the USA (and at PX outlets on bases abroad) since 1937. While it is almost certainly an accidental bonus, an image taken on this film - if properly cared for - will last more or less forever. This one deserves to.
(via Robert F.Dorr)

CHAPTER 3

North American F-100A Super Sabre

(Right) Rarely depicted in color is the short-tailed F-100A Super Sabre, the first production model to emerge from the North American factory. F-100A (53-1544), sporting an Air Training Command (ATC) tail emblem, is seen here in November 1954 on the very busy flight line at Nellis AFB near Las Vegas, where generations of F-86 and F-100 pilots received lead-in and gunnery training. The placement of the national insignia on the "Hun's" long nose and "U.S.AIR FORCE" on the tail was an early arrangement. By April 1955 203 examples of the F-100A had been produced for the bargain basement cost of $663,354; the final 39 machines had the earlier Pratt & Whitney J57-P-5 replaced by the more advanced J57-P-39 turbojet. Typically, an F-100A could fly out to a combat radius of 350 miles (525km) with internal fuel and a pair of 275-USgal. droptanks, leaving 15 minutes of loiter time over target to engage the enemy. This air-to-air profile was part of the standard training syllabus in the mid-1950s, but little used thereafter when the Super Sabre was relegated mainly to the fighter-bomber role.
(Curtis M.Burns)

(Left) This North American F-100A Super Sabre (53-1627) is en route from Clark Field in the Philippines to Tainan Airport, Taiwan, for delivery to the Chinese Nationalist Air Force as part of Operation Ghost Town. The Chinese F-100s were transported to the Philippines by sea, then flown to their new owners by American pilots like Captain Curtis N.Carley; the date is probably 9 May 1960, although such flights occured between 23 March and 23 June. This "Hun" is devoid of national insignia, but wears its serial number on the fin in standard 12-inch, and buzz number FW-627 in standard 15-inch numerals. The only other markings are the two jet engine warning lines painted round the rear fuselage (two, because the Pratt & Whitney J57 had two compression stages). The long pitot boom below the lower nose is capable of being folded for ease of storage. The dark rectangle behind the cockpit is a marker beacon.

The aircraft carries two 275-US gal. droptanks - standard for the period - but is not making use of its capability to carry additional fuel on the inboard pylons. Maximum fuel load for an F-100D was 2,545 US gallons, and its all-up combat weight around 28,040lbs (12,719kg). An unarmed F-100A was about 4,000lbs (1,814kg) lighter than the F-100D which Carley and the other pilots chosen for the ferry job were accustomed to flying; they had no previous experience of the A-model, and considered it an unbridled "hot rod of the skies". The F-100A, F-100C and F-100D were externally almost identical and had similar dimensions, although the D-model had a slightly larger wing.
(Curtis N.Carley)

When the Super Sabre entered service it was expected that most air refuelling would be carried out by the Boeing KB-50J Superfortress tanker. This F-100D in the colorful markings of the 356th Tactical Fighter Squadron, 354th Tactical Fighter Wing, is on a sortie from Aviano air base in Italy on 17 February 1959, and is linking up with KB-50J (47-0170). The process required the "Hun" pilot to plug his bent refuelling probe into the KB-50's drogue basket. Even with jet engines hanging under its outer wings the KB-50 was not really fast enough for efficient refuelling of jet fighters. When the Strategic Air Command was given responsibility for all US tanker operations, major

commands such as US Air
Forces in Europe (USAFE)
gave up their KB-50s and
from that point on KC-97
and KC-135 tankers
became more familiar.
(*Curtis M.Burns*)

(Above) "Rebel II" is a North American F-100D (55-3640) of the 510th Fighter-Bomber Sqn., 405th FB Wing at Clark Field in the Philippines in late 1959/early 1960. Mounted on the inboard pylon are 200-USgal. ferry tanks, normally fitted only when fuel was needed additional to the internal capacity and to the 275-USgal. droptanks carried further outboard. Note details of cockpit, nosewheel and leading edge slats. The major in the foreground is wearing G-pants over his sage green flight coveralls, and the USAF blue "overseas cap" (known universally to airmen by an unprintable term associated with the female anatomy). The enlisted man in the background seems to be an armorer, examining the access panel of the link compartment for the ammo belts feeding the four 20mm cannons.
(Curtis N.Carley)

(Top right) Night work on "Rebel II"; she is mounted on jacks, standard practice when boresighting the four M39E (formerly T-160) 20mm cannons.
(Curtis N.Carley)

(Right)
A good look at a silver "Hun" with blackened gun ports, open cockpit (note standard F-100 access ladder) and open access panels. F-100D Super Sabre (55-2894), pictured at Da Nang air base, South Vietnam, in August 1965, displays a variety of personal markings including "Kay Lynn", the name of pilot Donald Kilgus's wife. This aircraft is intriguing for other reasons, however. In April 1965 Kilgus mixed it up with a North Vietnamese MiG-17 only moments after MiGs had scored their first aerial victories of the Vietnam War - two Republic F-105D Thunderchiefs which they ambushed and shot out of the sky from "six o'clock". Kilgus chased the MiG through Vietnam's ever-present cloud cover to the sea, firing bursts of 20mm; he and wingman Ralph Havens believed for the rest of their lives that Kilgus scored a "righteous kill", but the US Air Force would only credit him with a "probable". Had Kilgus gotten full credit, it would have been the only aerial victory ever scored by an F-100. With Kilgus and Havens now deceased, it is unlikely that the record can ever be adjusted.
(Don Kilgus)

The cockpit of the F-100 was roomy and comfortable enough to prompt some pilots to dub this fighter the "Cadillac of the skies"; but it was also cluttered, especially for a single-engined aircraft. The primitive bomb lay-up system was positioned in front of the A-4 reflector gunsight combining plate (D in key drawing), hampering the pilot's forward vision. We showed several photos to a former F-100 pilot; but even he had trouble remembering the function of all the clocks, buttons, toggles and switches. Those we can identify are keyed in the drawing.
(Photo David W. Menard, drawing John Anastasio)

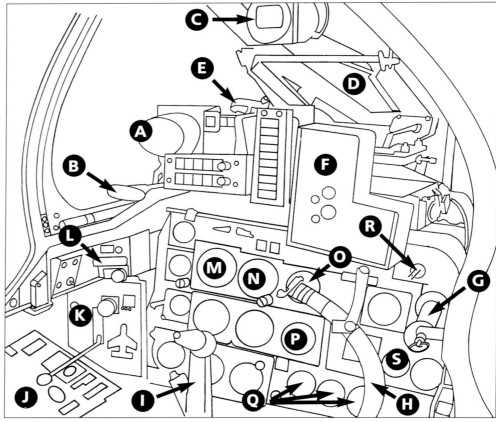

(A) Scope for radar homing and warning receiver; (B) drag chute handle; (C) Magnetic standby compass ("whiskey compass"); (D) Gunsight combining plate; (E) Dial for setting sight to correct wingspan of target aircraft; (F) Clear container for displaying let-down charts; (G) handle of knife for punching through jammed canopy; (H) Oxygen line; (I) Control column; (J) Console with controls for engine and fuel management, trim, weapons management, etc.; (K) Console of landing gear controls and lights; (L) Tail hook control; (M) Airspeed; (N) Attitude; (O) Rate of climb; (P) Altimeter; (Q) Fuel gauges; (R) Hydraulic pressure; (S) Temperature.

(Below & right) February 1966, Bien Hoa, South Vietnam: the fin shot of this F-100D Super Sabre (56-2954) appears to show battle damage, but this could be just routine maintenance - as usual, in the open revetment and with a minimum of tools and parts available. The outfit is the 416th TF Sqn.,"Silver Knights"; at this relatively early stage in the unfolding of the American tragedy in Vietnam the "Hun" is still in natural metal finish, though battered, dirty, and with the nose blackened by "corruption" from firing its 20mm cannons. The moderate ordnance load is also typical of the early days of the war. The 275-USgal. droptank, seen here on the outboard pylon though more typically carried inboard, had greater fuel capacity than was needed for most missions later in the war and was not much used. The 500lb HE bomb with detonation fuse on the inboard station was typically carried throughout the war. Note the bent refuelling boom extending from the starboard wing, with an intake plug and red flag to be removed before flight; by this date Pacific Air Forces no longer operated the KB-50J tankers which could have refuelled the "Hun" by this method.

The original fin of the F-100 was more than a foot shorter than we see here, and caused the loss of several aircraft due to roll-coupling, which turned a hard dive pull-out into an uncontrolled swerve to the right. One such crash on 12 October 1954 had claimed the life of North American test pilot George Welch, a 16-kill World War II ace who had made the first flight in a Super Sabre on 25 May 1953. At the upper rear of the taller fin are a radar beacon and command radio antenna, not fully visible, and position lights and a fuel vent outlet located at the rear of the fillet above the rudder. *(Don Kilgus)*

Maintenance conditions were no better in Vietnam than in Korea - and in summer it was even hotter and more humid. At close to 100 degrees F these 416th TF Sqn. armorers at Bien Hoa struggle with both linked belts of "twenty mike mike" shells and a 500lb bomb. The four M39E cannons could fire at a rate of 1,500 rounds per minute at a muzzle velocity of 3,300 feet per second; but each gun was equipped with only 200 rounds - not a particularly generous supply. Note the lifter used to raise the bomb for attachment to its shackles. This port inboard ordnance station - one of six hard points for bombs, rockets, missiles or fuel tanks - could have been used for the Mk 7, Mk 38, or Mk 43 nuclear store had the decision been taken to employ atomic weapons. The Super Sabre was also capable of carrying Martin AGM-12 Bullpup air-to-surface missiles, which did not perform well during early missions over North Vietnam. All single-seaters employed in Vietnam were of the F-100D model, first flown on 24 February 1956 by Dan Darnell. North American built 1,274 D-models; about 200 were lost in Vietnam. The last of the combat Super Sabres departed Vietnam in spring 1971 after nearly eight years of war service.
(Don Kilgus)

(Left) One item of ordnance employed on the F-100D early in the Vietnam War was the Air Force's SUU-17A bomblet canister, based on the standard canister for 19x2.75in. folding fin aircraft rockets (FFAR). Both types of ordnance were unguided, but highly effective against area targets such as troop concentrations. The canister was covered with a pointed frangible nose cone, not shown here, which reduced drag and which was jettisoned when rockets were fired or bomblets dispensed.
(Don Kilgus)

(Above) Two-seat F-100F Super Sabre on a combat mission over Vietnam. Conceived as a "lead-in" trainer rather than a warplane, the F-model was three feet longer than the F-100D; its combat weight was up to 31,413lbs (12,719kg) and its fuel capacity down to 2,259 US gallons. It was armed with two M39E 20mm cannons, and all armament controls were grouped in the front cockpit; a student or observer took the rear seat. Apart from the cockpit it was virtually identical to the D-model; and because its performance was exactly the same, with the exception of a slightly reduced combat radius due to the sacrifice of 186 gallons of fuel capacity, it was in fact employed in Vietnam almost immediately. Later in the war the F-100F was used on "Misty FAC" (forward air control) missions; and was adapted to become the first "Wild Weasel" for suppression of North Vietnamese ground-to-air defenses.
(Don Kilgus)

Bien Hoa, early 1966: like many fighters of its era the "Hun" was designed with a break just behind the wing trailing edge, making it possible to remove the tail section and rear fuselage. This "Silver Knights" F-100D truly affords us a glimpse "inside a great jet" - the rear section of the 16,950lb (7,693kg) thrust Pratt & Whitney J57-P-21A axial flow turbojet with the tail removed; and being pulled for maintenance or repair, using a typically low-tech, muscle-and-sweat rig. The J57 was about as user-friendly as any jet power plant of the period, but was far more difficult to reach and work on than the engines of today's warplanes. Note the afterburner exhaust nozzle; with burner the thrust was increased to 17,000lbs (7,711kg).
(Don Kilgus)

(Above) Later in 1966 the TO 114 camouflage scheme began to appear on Super Sabres in Vietnam. "Hun" pilots flew closer to the Viet Cong than anyone else in fast jets, and suffered many of their losses from small arms fire while supporting US and ARVN troops in battle.
(Don Kilgus)

(Top right) A hard-charging F-100 crew chief struggles to earn a living during a deployment to Korea at the time of the 1968 *Pueblo* crisis. Like a scaly reptile, the F-100 was covered almost from nose to tail with inlets, panels, and doors for access to its interior systems. Here, in front of the windshield we see the raised hood above the ammunition boxes for the four 20mm cannons; and the open panel behind the rear segment of the canopy gives access to the rear electrical bay situated just behind the pilot.
(Don Porter)

(Right) Close-up comparison of two very different jet fighters at Bien Hoa in November 1968. In the foreground, Major Jan Holley, an advisor to the RVNAF 52nd Fighter Squadron, sits (wearing typical flying helmet and garb of the period) in the cockpit of a Northrop F-5C Freedom Fighter (65-10515), the lightweight fighter built primarily for export. Beyond is an F-100D Super Sabre assigned to Capt.Gene Westback, his name typically marked on a red field on the canopy rails. Pilots claim that the TO 114 camouflage paint job took

about 20mph off the speed of the aircraft and made its performance less smooth. Standard practice was to use afterburner for take-off; one pilot with a remarkable 2,000 hours in his Form Five (flight log book) says:"I never made a take-off in the One Hundred without using burner".
(Curtis M.Burns)

(Left) A good view of the TO 114 camouflage, modelled by an F-100D (56-3324) of the 90th TF Sqn., 3rd TF Group, up from Bien Hoa air base in January 1969; the pair of 275-USgal. droptanks are a holdover from the earliest days of F-100 operations. The practice of putting the "last three" of the serial number behind the lip of the air intake was adopted late in the Super Sabre's career, and was intended to save time for hard-pressed ground crews working on a busy flight line. The CB tail code identifies the 3rd TF Group. The blade on the spine of the aircraft for a radar warning receiver is also a late Vietnam War addition. The pilot's name appears in white in a white-bordered blue panel just below the canopy.
(Curtis M.Burns)

(Bottom left, below & overleaf) Close-up details of a very late F-100D (55-2939) in South East Asia, circa 1970, by which time the aircraft was deemed obsolescent. The 416th TF Sqn. wore the SE tail code while operating the Super Sabre; the unit later graduated to the F-4 Phantom.
(Donald L Jay)

Further angles on F-100D (55-2939) of the 416th TF Sqn. Note the detail of the (seldom used) air refuelling probe, with its integral light for night operations. Viewed up close the canopy is remarkably clean. Pilots had confidence in their ejection seat, which nearly always worked as advertised. (Donald L Jay)

CAUTION WHEN CLOSING STRONG SPRING

RESCUE

PUSH

CHAPTER 4

McDonnell F-4 Phantom

(Above) McDonnell F-4E Phantom (68-306) in a revetment during a visit to Ubon, Thailand, on 24 August 1971. The aircraft belongs to the 421st Tactical Fighter Squadron (LC tail code), 366th Tactical Fighter Wing "Gunfighters" stationed at Da Nang, South Vietnam. The Phantom wears standard TO 114 camouflage (named for the Technical Order which prescribed the colors). The bullet-shaped fairing for a radar warning receiver at the top rear of the fin was a feature which arrived rather late in the Phantom's long career, and was intended for the AN/APS-107 system, which was never installed in most aircraft. The relatively wide track of the landing gear is evident here.
(via Robert F.Dorr)

(Top right) Ubon, Thailand, spring 1967 - and the 8th Tactical Fighter Wing "Wolfpack", commanded by World War II ace Col.Robin Olds, is committed to intense combat against North Vietnam's MiGs, SAM missiles and anti-aircraft artillery ("triple A"). This F-4C, crewed by Major Truman Spangrud and 1st Lt. Roger Iversen of the 435th TF Sqn., gives us a good impression of a Phantom carrying centerline fuel tanks and Mk 80 series bombs on the pylons; note also ladder and nosewheel detail evident from this angle. Both men will use the same ladder to mount the aircraft; the back-seater - at this date a fully-fledged pilot, though later to be a navigator trained as a weapons systems officer (WSO) - will need agility and good balance to contort himself back into his office under the rear canopy.
(Truman Spangrud)

(Right) Cam Ranh Bay, Vietnam, 1967: this F-4C Phantom (63-7507) belongs to the 559th TF Sqn., 12th TF Wing. This radome shape with infra-red sensor fairing beneath the nose is typical of F-4Cs and F-4Ds of the Vietnam era, but was not used for any purpose on most Air Force aircraft until the introduction in the late 1960s of the Bendix AN/APS-107 RHAWS (radar homing and warning system). This Phantom carries a centerline bomb load apparently consisting of standard 750lb (340kg) Mk 83 bombs, plus a pair of cluster bomb canisters hanging from racks on the inboard pylons, plus standard droptanks outboard. Vietnam-era TO 114 camouflage employed three colors on the upper surfaces - forest green (Federal Standard 34079), medium green (FS 34102) and tan brown (FS 30219); the underside was painted very pale gray (FS36622). This particular Phantom survived the war and later served at Edwards AFB, California, in the late 1970s and 1980s, during which period it was the oldest Phantom still flying.
(Richard Kamm)

A brief but colorful ceremony (involving a smoke grenade intended as a rescue signal) on the flight line at Ubon marks the completion of 1st Lt. Steve Mosier's 100th combat mission over North Vietnam. This provides a close-up glimpse of "Carolyn", an F-4D (66-7767) of the 433rd TF Sqn. (FG tail code), 8th TF Wing "Wolfpack". Painting of the aircraft nickname on the nose cooling scoop is unusual. The impression of tourists in attendance is misleading - the guy in Bermuda shorts is one of the airmen at the base. (Steve Mosier)

(Above) A heavily bomb-laden F-4D (66-7730) of the 390th TF Sqn. (LF tail code), 366th TF Wing "Gunfighters" out of Da Nang, tops up with fuel from a KC-135 Stratotanker while en route to a target in South Vietnam in about February 1970. Air refuelling during combat missions was first tried during the Korean War; in Vietnam it became routine. The extended fuses on the Mk 80 series bombs identify so-called "daisy-cutters"; these detonated a short distance above the ground, spreading blast and fragmentation further than a surface detonation, whose effect might be degraded by the bomb digging itself into the soft soil. The blue squadron color is just visible at the top of the rudder and in the area under the canopy, where the crew's names are painted.
(via Robert F.Dorr)

(Top right) February 1970: a "bombing halt" is keeping American warplanes out of the skies of North Vietnam, but in the South the war continues unabated. Loaded for bear, an F-4D Phantom (65-0771) of the 390th TF Sqn. (the LF tail code just visible cut by the left edge of the image) trundles out towards the end of the runway at Da Nang for a combat mission. The very large 680-USgal. tank under the centerline compensates for the absence of tanks on the underwing pylons, and may be a sign of a long range or long "loiter" mission. Note the Mk 80 series bombs, and the white-painted AIM-7E Sparrow III radar-guided missiles.
(via Robert F.Dorr)

(Right) F-4D Phantom (66-8802) of the 435th TF Sqn., 8th TF Wing "Wolfpack", on a combat mission from Ubon, Thailand in 1972. This aircraft carries a dorsal "towel rack" antenna for AR/ARN-92 LORAN (long range navigation) equipment, enabling the Phantom to guide itself and perhaps a flight of warplanes on a "Sky Spot" radar guided bombing mission. F-4D Phantoms were originally delivered with "slick" bullet-shaped radomes, to which the chin mounted infra-red sensor housing seen here was later added. This ship also carries an AN/ALQ-119 electronic countermeasures pod.
(via Robert F.Dorr)

(Above) The JV tail code identifies an F-4E (68-0322) of the 469th TF Sqn., 388th TF Wing, here flying a combat mssion out of Korat, Thailand in March 1972. The F-4E was the only Phantom model equipped with a nose-mounted General Electric M61A Vulcan rotary cannon, and 639 rounds. The cannon housing shown here - the "shark's chin" - is a configuration retro-fitted after the aircraft entered service; earlier, the housing had a more rounded front. The narrow light-coloured stripes on the fuselage and fin are nighttime formation-keeping lights - known to some aviators as "slime lights" because of their sickly green color when illuminated. The white triangular shape visible behind the wing trailing edge is the fin of an AIM-7E Sparrow semi-active radar missile on its mounting under the fuselage. *(via Robert F.Dorr)*

(Left & below) A big, sturdy warplane with multi-mission capabilities, the F-4 Phantom was able to haul just about any bomb, rocket or missile in the USAF's Vietnam inventory; and in a war which saw the United States dropping five times as many bombs as in World War II, that gave plenty of choice. In practice, however, the majority of the ordnance dropped by Phantoms came from the Mk 80 bomb series. This family of gravity bombs, introduced during the Vietnam War, differed from one another only in size and displacement. The standard weapons were the Mk 81 of 250lbs (113kg); the Mk 82 of 500lbs (227kg); the Mk 83 of 1,000lbs (454kg), used only by the US Navy; and the Mk 84 of 2,000lbs (907kg); there was also the M117 of 750lbs (340kg). The weights of these GP (general purpose) bombs - still in widespread use today - represented roughly one-half casing and one-half explosive. A number of different fin configurations could be attached to them, including the conical LDGP (low drag general purpose), and the high drag Snake-eye type used to retard a bomb's trajectory. The Mk 80 family served as the basis for the development of the "smart" precision guided munitions which were introduced late in the war, and which changed the shape of the air-ground battle forever.

Here we see a typical "escort load" as carried by an F-4 in SE Asia for the purpose of accompanying reconnaissance or rescue aircraft into harm's way. Three Mk 82 bombs are mounted astride a triple ejector rack on the outer ordnance station each side, with four CBU-24 cluster bombs on the inboard station; the CBUs are effective against "soft" targets such as AA gun positions and troop concentrations. The close-up shot shows a live 500lb HE Mk 82 carried by an F-4E; 34th TF Sqn., Korat, Thailand, 1972.
(Donald L Jay)

(Left) This sharkmouth F-4E of the 388th TF Wing at Korat carries a SEAD (suppression of enemy aerial defenses) ordnance load, for neutralising the source of any ground fire encountered by US reconnaissance aircraft. On the centerline and inboard pylons are the CBU-24/29 cluster bomb units which were the prime weapons against gun positions. Note the covering for the small pitot tube in the nose; and below it the distinctive cannon housing - the F-4E was the first Phantom with an internal gun, and carried a smaller radar dish than previous models in order to accomodate the 20mm M61A1 "Gatling" cannon. This photo was taken in September 1973; America's combat involvement in Vietnam was over, but US aircraft still flew recce missions and were ready for anything they might come up against. Phantoms had been heavily committed to combat since 1965, throughout the whole period of US commitment in SE Asia; they remained until the final withdrawal of US forces, and returned briefly during the evacuation of Saigon in April 1975.
(Donald L Jay)

(Bottom left) An F-4C Phantom with the data block correctly applied on the side of the fuselage beneath the cockpit - almost. . . Type letter F, model number 4, and series letter C give us an F-4C fighter - the fourth principal type of fighter in use since the current designation system was adopted on 1 October 1962. The number 19 identifies the production block in which the aircraft was manufactured; and the letters MC identify the manufacturer as the McDonnell Aircraft Company in St.Louis, Missouri. Before, during, and long after the production run of the Phantom this company remained in existence, known in the 1950s as MAC and in the 1970s as McAir - thus the Phantom was always a McDonnell, rather than a McDonnell Douglas aircraft. The serial number tells us that this was the 7,592nd airplane or missile to be contracted for (though not necessarily built) with fiscal year 1963 funds. The A suffix to the serial is incorrect: it is supposed to tell us that this is a regular Air Force aircraft - as distinguished from an Air National Guard ship, which should bear a G suffix. In fact this Phantom did belong to the 199th Fighter-Interceptor Squadron, 154th FI Group, Hawaii ANG when it was photographed at Hickam Field in August 1977. The wedge-shaped device is a retro-fitted rear view mirror for the back-seater.
(Robert F.Dorr)

(Below) A post-Vietnam portrait of a MiG-killer: F-4C (63-7584) was photographed in June 1975 - just days after the evacuation of Saigon - when serving with Brig. Gen. Fred A. Haeffner's 58th Tactical Fighter Training Wing at Luke AFB, Arizona. She bears recent stencilling, sharper and brighter than that sported by aircraft in SE Asia - judging from the paint can and spray containers standing around, "Fat Freddie's" personal Phantom has received very recent attention. The red stars represent one aerial victory plus one "probable" racked up by Haeffner. The pilot's name is stencilled on the front canopy rail, the crew chief's on the rear rail. The data block beneath the cockpit should identify this aircraft as an F-4C-19-MC - "MC-F-4C" is an error. This angle illustrates the difficulty faced by the WSO in climbing into and out of the back seat using the pilot's ladder; standard procedure was to step over the air intake, plant a boot in the middle of the ejection seat, and ease on down.
(Michael Grove)

(Above) Tail close-up of F-4C (63-7592) at Hickam Field, August 1977. The Vietnam era colors survive, but not the unit tail code. Note too the retention of the tail hook by Air Force Phantoms. See again the football-shaped blister on the fin cap, intended to accomodate the AN/APS-107 radar homing and warning system which was never installed in most aircraft; and the slanted "slime light" which helped crews keep formation and distance at night. The drag chute compartment was located in the tail cone, beneath a rearward pointing probe which housed a static discharge vent. The Phantom's slotted stabilator had 23.25 degrees of anhedral.
(Robert F.Dorr)

(Top right) Post-Vietnam colors. In the 1980s most USAF tactical aircraft changed to the vomit-like "lizard green", also known as "Europe One" - a camouflage scheme ideal for hiding a Phantom in a clump of trees, but not very aesthetic. Led by the wing commander's aircraft of the 37th TF Wing, these McDonnell F-4G Advanced Wild Weasels are on a sortie over the Sierra Madre in California, flying from George Air Force Base at Victorville. The other three Phantoms each bear the color of one of the wing's squadrons, and highlighted designations on the fins. The F-4G is distinguishable from other Phantoms by the "hump" atop its vertical tail which carries electronics equipment related to its job of attacking surface-to-air missile radar sites. This Wild Weasel mission emerged from the Vietnam War; F-4Gs fought in this role in the Gulf War of 1991, but have since been retired, and the USAF has no plans for a "dedicated Weasel" in the future. The ordnance carried here is the AGM-78 Standard ARM anti-radar missile, since replaced by the AGM-88 HARM.
(USAF via R.J.Mills)

(Right) A "wing king" Phantom: this reconnaissance RF-4C (69-0363) of the 363rd Tactical Reconnaissance Wing, stationed at Shaw AFB, South Carolina, was photographed while on a visit to Andrews AFB, Maryland, in May 1977. It is painted in TO 114 colors, much as it might have appeared in the war zone a couple of years earlier. Assigned to the wing commander, this Phantom displays the colors of all three flying squadrons of the 363rd on the fin cap; the wing designation, corresponding to the "last three" of the serial, highlighted; the wing's JO tail code; and the Tactical Air Command badge.
(Robert F.Dorr)

(Left) Close-up of a MiG-killer. F-4C (63-7647) was serving with the 199th FI Sqn., 154th FI Wing, Hawaii ANG when photogaphed at Hickam Field in July 1984. She still wears Vietnam era colors, except that the very pale grey underside has been replaced by "wraparound" camouflage. Note that the pilot's ladder has an asymmetrical left hand grip at the top; most pilots and "wizzos" are right handed, but the ladder is designed for their direction of entry into the aircraft from the traditional starboard side. One of the red stars commemorates a MiG-17 shot down with an AIM-7E Sparrow AAM on

5 June 1967 by Capt.Richard M. Pascoe and 1st Lt. Norman E.Wells, when this Phantom was serving with the 555th TF Sqn." Triple Nickel", 8th TF Wing.
(Robert F.Dorr)

(Above) Another double MiG-killer enjoying a second career with the Air National Guard in sunny Hawaii in July 1984, F-4C (63-7676) "Noio" displays the telescoping integral pilot's access ladder. Although the fiction was widely published that the F-4C and F-4D had different sized radomes, both in fact had the size illustrated here, although some were delivered from the factory without the chin pod for an infra-red sensor (which was usually empty, anyway). On the side of the air intake, low down, is painted a check mark, the number 6, and the word "JUVATS". "Check six" is pilot jargon for "watch your six

o'clock position" - the area dead astern; and JUVATS is shorthand for the Latin motto (translating roughly as "Fortune favours the bold") of the 80th TF Sqn. at Kunsan, Korea, indicating that this Hawaii Guardsman has recently been received from the 80th.
(Robert F.Dorr)

(Below) This "air defense gray" scheme was sported by a few F-4s in fighter-interception squadrons during the 1970s and 1980s, at a time when most other Phantoms were still wearing Vietnam style camouflage. This is an F-4C (63-7505) of the 199th FI Sqn., Hawaii ANG at Hickam in August 1984. Later in the 1980s this unit converted to the F-15 Eagle. (Robert F.Dorr)

(Bottom) Spot the deliberate mistake. . . In the 1980s, when shifting from Vietnam era TO 114 camouflage to low visibility gray, the Hawaii ANG's paint crew showed excessive zeal in applying the toned-down US national insignia in two locations on this F-4C, which may have been the only Phantom ever to wear the stars and bars on both forward and aft fuselage. Legend has it that "Pobo" (64-0792) flew around in these incorrect markings for some months. When photographed in July 1984, however, this aircraft was going nowhere: note that part of the left wing inboard leading edge flap has been removed for repair. The fin band identifies the 199th Fighter-Interceptor Squadron. (Robert F.Dorr)

(Right) "Alae Ula", a Hawaii ANG F-4C Phantom (63-7647) photographed at Hickam in July 1984 still proudly sporting on the air intake splitter plate the stars marking two MiG kills in Vietnam. Note also the details of the radome, formation light and nosewheel shown to advantage by the brilliant tropical sun. Note the air intake cover intended to protect the General Electric J79 turbojet from FOD (foreign object damage) while running up. On this Phantom the name of the AC (aircraft commander) and WSO (weapons systems officer) are stencilled in brilliant yellow against the camouflage scheme. (Robert F.Dorr)

(Right) Low visibility gray. The last operational Phantoms went out (visually) not with a bang but a whimper: no more Vietnam sand-and-spinach, no more Cold War lizard green, just a ghostly fading away into this scheme designed to hide them not on the ground but in flight. During exercise Sentry Wolverine on 17 June 1987 this F-4D (65-710) of the 184th TF Group, Kansas ANG - the "Jayhawks" - has closed on a KC-135 tanker for a fuel top-up. Several variations on the drab gray color scheme appeared on Phantoms; this particular design is sometimes called "Egypt One", as the Egyptian Air Force was the first to use it. Carrying only a device which simulates an AIM-9 Sidewinder missile for training purposes, this Wichita-based crew has "buttoned up", leaving no sign of the air refuelling receptacle on the spine of their fighter-bomber as they pose for a classic portrait. Their unit later flew F-16s before converting to the B-1B Lancer. (David F.Brown)

TO RELEASE

FWD CANOPY

(Left & above) The cockpit shot, showing the pilot's workplace in a veteran F-4S, would be familiar to several generations of fliers. There were no CRTs, HOTAS or HUD for aviators of the Vietnam era: just a dim luminescent scope for the AN/AWG-10A fire control radar, conventional artificial horizon and altimeter gauges, and an optical gunsight on top of the cockpit coaming. This particular Phantom would have left St.Louis in the late 1960s as an F-4J; the upgrading modifications performed a decade later by the Naval Aircraft Rework Facility have left the cockpit relatively unchanged.

The ejection seat close-up features an F-4E DMAS (Digital Modular Avionics System) of the Air Force Reserve's 704th Tactical Fighter Sqn. at Bergstrom AFB in 1989; the new digital modular nav-attack system enabled the Phantom to deliver "smart" munitions such as the AGM-65 Maverick. The seat is the Martin-Baker Mk H7, initiated by pulling either a D-ring between the pilot's legs or the striped O-ring behind his head; it promises safe ejection and relatively happy landings at anything down to zero altitude and zero speed.
(Joe Cupido)

(Above) It is 1988, and the end of an era is approaching. This in-flight portrait of F-4Ds of the District of Columbia's 121st TF Sqn., 113th TF Wing stationed at Andrews AFB near Washington was in fact taken during a temporary deployment to Norway. With the extraordinary advances in navigation and weapons control systems over recent years, it is not just the F-4D which is approaching retirement: the GIB ("guy in back", or weapons system officer) is also about to be put out to pasture. By the mid-1990s the US Air Force will have only one fighter-bomber left in inventory with a "back-seater" - the F-15E Strike Eagle from the same manufacturer, serving in relatively modest numbers. *(DC ANG/ Thomas F.Evans)*

(Top right) For many years this attractive red and white Phantom was a fixture at the US Air Force Flight Test Center (AFFTC) at Edwards AFB, California, where generations of test pilots have pushed at the outside of the envelope. F-4C (63-7654) of the 6512thTest Squadron, photographed in June 1990, is wearing everything, including an Air Force Systems Command badge on the fin. The general use of the "U.S.AIR FORCE" in foot-high letters on the side of the fuselage began in early 1953, but had been discontinued on tactical aircraft long before this portrait was taken. Edwards was one of the very last locations to have Phantoms in operation. *(via Robert F.Dorr)*

(Right) A start-up by a McDonnell F-4E Phantom (68-0351) of the 110th TF Sqn., 131st TF Wing, Missouri ANG in January 1990. Unlike almost every other Phantom this one has a single-piece windshield, an item the manufacturer hoped to sell to the ANG in the 1980s. Note the sharkmouth, squadron insignia, and tail markings; and the F-4E model's nosewheel, tucked neatly behind the housing for the 20mm cannon, with the "last three" of the aircraft serial marked on the door. The 110th TF Sqn. was directly across the field at Lambert-St.Louis Municipal Airport from the McDonnell Aircraft factory which turned out 5,067 Phantoms between 1959 and 1979. *(Robert F.Dorr)*

(Above) The Missouri Air National Guard going to work: the crew are preparing to crank up F-4E (68-0351) - note again the rare one-piece windshield. The crew chief is poised somewhat precariously as he sees the back-seater into his place of business. The crew's parachutes are, of course, stowed as part of the ejection seat installation rather than being strapped on the body as in days of yore; the slick area on the WSO's harness is just a pad. When the crew are strapped in the crew chief will return to the ground, watching out for possible problems during engine start and communicating with the pilot via a plug-in intercom lead. *(Robert F.Dorr)*

(Top right) "Wild Weasel, Wild Weasel, they call me by name. . ." - so begins a bawdy Air Force song about taking on the bad guys, most of which is unfit to appear in this family publication. Two F-4G Advanced Wild Weasels deployed from the 52nd TF Wing at Spangdahlem air base, Germany, fly a combat mission during Operation Desert Storm, 1991. They sport the drab gray paint scheme which became standard in the 1990s, and carry AGM-88 HARMs (High-speed Anti-Radiation Missiles) and jamming pods. The USAF had been on the verge of retiring its fleet of F-4Gs - their only warplane dedicated to attacking enemy air defense missile sites - when the Persian Gulf confict broke out. Saddam Hussein's stunning miscalculation allowed the much-loved Phantom the privilege, rare for combat aircraft types, of going out in one last blaze of operational glory. *(USAF)*

(Right) The end of the trail: on 24 January 1992 an F-4G Advanced Wild Weasel (69-7293) from the 52nd TF Wing at Spangdahlem, Germany (SP tail code), sits on the line at the Air Force storage and disposition center at Tucson, Arizona, awaiting a slot in the "boneyard" after being withdrawn from service. The Phantom seems still to carry the long range ferry fuel tanks which enabled it to get from Germany to Davis-Monthan AFB in the sunny southwest. Details of the F-4G's distinctive nose and tail configuration are evident here. After this valedictory portrait was snapped the Phantom was sprayed for corrosion, swabbed with ablative, assigned a storage number, and placed in outdoor storage - in theory capable of being recalled for duty at future need, but unlikely ever to be so. Some may feel that the USAF's decision to do without a Wild Weasel capability in future flies in the face of the lessons of the Gulf War. *(Robert F.Dorr)*

CHAPTER 5
Lockheed F-16 Fighting Falcon

(Left) F-16C block 40/42 Fighting Falcon (89-2067) of the 68th Fighter Squadron, 347th Fighter Wing at Moody AFB, Georgia (MY tail code), in flight with a typical load for an air-to-ground training mission. Inert AIM-9 Sidewinder training missiles are carried at the wingtips (ordnance stations 1 and 9); small blue BDU-33 practice bombs on three of the four outboard pylons (stations 2, 3 and 8); a Westinghouse AN/ALQ-119 electronic suppression system jamming pod, visible here only as a black outline under the far wing station (station 7); and two 370-USgal. (1,378 litre) fuel tanks on stations 4 and 6. The blade antenna just ahead of the fin distinguishes F-16C/D models from F-16A/B aircraft. At the time of this portrait the 347th was a typical fighter wing; it has since become a composite wing operating several types of warplanes, still including the Fighting Falcon.
(*Michael Hagerty*)

(Below left & right) On the ramp at Sandston Field (Richmond International Airport), November 1994: a close-up and general view of Lockheed F-16C block 30F Fighting Falcon (86-0229) of the 149th Fighter Sqn., 192nd Fighter Group, Virginia Air National Guard - courtesy of which proud unit (previously equipped over the years with F-47s, F-84s, F-105s and A-7Ds) the reader is about to enjoy a detailed walk-around. The austere VA tail code and "Virginia" logo on the fin reflect 1990s political correctness. In June 1992, while the squadron still had Vought A-7D Corsair IIs, Virginia Governor L. Douglas Wilder banned the use of the Army of Northern Virginia battle flag which had previously been the trademark of this outfit. In 1996 the Virginia Guardsmen at Sandston Field were scheduled to begin operating a new recconnaissance pod with their F-16Cs.
(*Robert F.Dorr*)

(Left) Moving back along Viper (86-0229), we find that somebody in maintenance has used a spraycan and stencil to apply the US Air Force serial number of a Boeing AGM-86B air-launched cruise missile (84-0229) - though photos show that the serial is correctly marked on the tail. This view silhouettes the pilot's HUD (Head-Up Display). F-16 canopies showing this dark, slightly gold-tinted color in contrast to a completely clear appearance have been treated with RAM (radar-absorbent material). The RESCUE sign points back to the best spot to extricate a pilot in a ground mishap. (Robert F.Dorr)

(Bottom left) The nose and radome of F-16C (86-0229). The red-flagged protective covers have been removed from the nose pitot tube, and from the spike-shaped object protruding at 90 degrees just forward of the DO NOT PAINT warning on the radome - this is an angle of attack (AOA) sensor. On the other (starboard) side an identical AOA spike appears along with a back-up, L-shaped AOA probe known to crews as the "dog pecker". The oblong gray bulge just forward of the RESCUE sign is the location where radar warning receiver (RWR) sensors were originally mounted; on many F-16Cs and Ds (but not F-16As and Bs) these have been moved to the outboard sector of the wing leading edge flap. Although no longer serving a purpose the bulge has never been flattened over - it has no aerodynamic effect.
(Robert F.Dorr)

(Right) Nose gear and rear cockpit area. Beneath the inverted triangle which warns of the explosive qualities of the Douglas-designed, Weber-built ACES II (Advanced Concept Ejection Seat) we see the gun port with twin gas vents for the muzzle of the Martin Marietta (formerly General Electric) M61A1 Vulcan 20mm rotary cannon, located at the point where the port wing blends into the fuselage. The cannon uses PGU-28 shells, which have more mass and are aerodynamically more efficient than the earlier Mk 56 ammunition which is now gradually being replaced throughout the fleet; 511 rounds are carried. The red light on the port side of the intake is for station keeping, and an identical blue light appears on the starboard side; pilots say that this is less effective for night and bad weather formation flying than the "slime light" strips seen on McDonnell fighters.
(Robert F.Dorr)

(Right) A glance inside the F-16C block 30F. The most immediately noticeable feature is the unique sidestick controller replacing the time-honoured centrally mounted control column; left-handed pilots say that they have no problem adjusting to this. The flat plate of the HUD is not readily apparent here, being side-on at this angle. Note the PULL TO EJECT legend on the handle between the pilot's knees.
(Robert F.Dorr)

(Below, right top & bottom) Undercarriage and belly details of F-16C (86-0229): the nosewheel (whose large door is peculiar to the F-16C/D models), the lower mid-section, and the main landing gear.
Instead of a 300-USgal. centerline fuel "bag" this Viper is fitted with a jury-rigged "travel pod" which gives the pilot space for a suitcase or other luggage. Whether fuel or clean shirts, the centerline store hangs beneath a rectangular pylon beam. All F-16s also carry two 360-USgal. tanks on inboard stations 4 and 6; and have the capability to carry two 610-US gal. tanks, though the latter are not stocked. Behind the main gear - a complex, inwards-folding assembly incorporating a taxi light - are two ventral fins for lateral stability, forming a slight V-shape below the rear fuselage.
(Robert F.Dorr)

F-16C block 30F (86-0266) in the hangar at Sandston in November 1994 for inspection pending transfer from Virginia to the Indiana ANG at Terre Haute. The cockpit canopy and ejection seat have been removed, and access panels opened to allow maintenance personnel to check and replace LRUs (line replaceable units) or modular avionics components - the modular design of the aircraft's innards reduces the number of times the F-16 has to be sent back to the depot for major maintenance. The nose radome is swung towards the camera, concealing some elements of the Westinghouse APG-66 radar. Note the blue tarpaulin intake cover; the thick black electrical connector lead used when running up systems during maintenance; and the ladder - normally used on the other (left) side of the aircraft.

The ladder is no small investment for the F-16 operator, and two-seat F-16B and D models require two different types of ladder matching the different fuselage contours. One cash-strapped customer - Denmark - forewent the purchase of crew ladders altogether, and Danish jet jocks climb into their Vipers up a 40-year-old maintenance stand built for the F-100D Super Sabre.

One general point illustrated by these shots is the immaculate cleanliness typical of an Air National Guard installation; in sharp contrast to the regulars, Guardsmen could safely perform surgery or eat their dinners on the hangar floor. (Robert F.Dorr)

(Top) The F-16 ADF (Air Defense Fighter) emerged as a major program in the late 1980s while the Cold War was still thawing - a modification of existing block 15 aircraft for strategic defense of North America against bombers and cruise missiles. ADFs were ordered in October 1986 to replace F-106s, F-4s and F-15s in the interceptor role with ANG squadrons. This nose close-up of an interceptor shows appendages found on all Fighting Falcons, plus some which are unique to the 273 airframes converted in 1989-90 to F-16A/B block 15 ADFs. Readily visible is the Grimes-built ID light found on ADFs and some European F-16s; not a searchlight, since it illuminates a target only after the target is found, this is fitted on the left side of the aircraft - which is odd, since it thus illuminates the target's right (co-pilot's) side. Greek F-16C block 30

fighters are the only Vipers which have the ID light on their right side. The ID light is canted 70 degrees to the left of forward and 10 degrees up; this annoys pilots, since the radar gimbal limit for the AN/APG-66(V)1 air-intercept radar is 60 degrees - meaning that the F-16 ADF cannot aim both ID light and radar at a target simultaneously.

Spikes protruding from the radome are standard AOA sensors. The darker gray fairing above the ID light houses the radar warning receiver (which has been moved to the wing leading edge flap on later F-16C/Ds); the twin blade antennas pointing downward beneath the radome are for the radio system and are found on all F-16s. A row of four blade antennas in front of the windshield, matched by four more (invisible from this angle) beneath the air intake, are for the Teledyne/E-Systems AN/APX-109 Mk XII AIFF

(Advanced Identification, Friend or Foe) system unique to the ADF and not approved by the US Government for export.
(Robert F.Dorr)

(Right top & bottom)
Big friend and little friend, up from Andrews AFB, Maryland, in November 1989. The Boeing C-22A (83-4615), formerly an airliner belonging to National Airlines and Pan American World Airways, is one of only four Boeing 727s employed by the Air National Guard, and is piloted here by Major Mark Hetterman. The F-16A block 10 (80-0250) is flown by Lt.Col.Vince Shiban, commander of the District of Columbia ANG's 121st TF Sqn., although it actually wears the markings of the 13th TF Wing commander, Brig.Gen.Russell C.Davis. This Fighting Falcon - long since replaced in its Guard squadron by the newer F-16C - illustrates the smaller horizontal tail found only on the earliest F-16s; today the only serving examples of this configuration are in Israel.
(Robert F.Dorr)

(Left) Cockpit of an F-16A/B block 15 ADF belonging to the "Jersey Devils" - 119th Fighter Sqn., 177th Fighter Group, New Jersey ANG, at Atlantic City in November 1994, when these aircraft were in progress of being replaced by newer F-16C block 25s. The head-up display on a flat plate tilted back towards the pilot gives him electronic cues about flight performance which enable him to fly in the HOTAS mode (Hands On Throttle And Stick) with his eyes up. Ahead of the ejection handle is the pilot's square shaped multifunction display. *(Robert F.Dorr)*

(Bottom left) This rear view of an F-16B block 15 ADF shows the "all flying" horizontal tail surface, enlarged to this size on all aircraft from block 15 onwards. Close against the rear fuselage and in the open up-and-down position are the speed brakes common to all F-16s. The exhaust petals are typical of all F-16s powered by Pratt & Whitney F100-P-100 turbofan engines, i.e. Fighting Falcons in production blocks 1, 5, 10, 15, 25, 32 and 42. The long, slender canopy braced in the open position is the only clue from this angle that we are looking at a two-seat

F-16B. The two-seat models - F-16B and D in all production blocks - retain the wing and fuselage dimensions of the single-seater but sacrifice 1,500lbs (580kg) of fuel, and hence endurance. The F-16B block 15 ADF differs from the F-16A block 15 ADF in not having an HF radio set; there is thus no bulge at the base of its fin where actuators were relocated in order to make room for the HF antenna on A-models in this series. *(Robert F.Dorr)*

(Below) As mentioned above, one prominent feature found on single-seat F-16A ADFs but not two-seat F-16Bs is the

Bendix-King AN/ARC-200 high frequency single-sideband radio, deemed highly effective for the long range transmissions required during an interception fly-out. Because of the addition of the HF radio's antenna on the leading edge of the fin, a pair of hydraulic actuators for rudder operation previously mounted one above the other were moved down, forward and side-by-side, giving the base of the fin this distinctive bulge. It is a widespread misconception that the bulge itself is an HF antenna. *(Robert F.Dorr)*

(**Left**) A Fighting Falcon interceptor at the ready, with the pilot alert and ground technicians standing by for launch on an intercept mission. This F-16A block 15 ADF belongs to the 119th Fighter Group "Happy Hooligans", North Dakota ANG, normally stationed at Hector Field in Fargo but seen here competing in the annual William Tell air-to-air shooting competition at Tyndall AFB, Florida, in late 1994. This angle shows the two rows of four blade antennas - ahead of the cockpit and under the intake - for the interceptor's AN/APX-109 Mk XII AIFF system. Note the single-piece canopy, and canted ejection seat position; the HUD at the top of the instrument console; the dark teardrop fairing of the RWR (Radar Warning Receiver); and the L-shaped starboard side angle of attack probe beneath the AOA sensor spike.
(*David F.Brown*)

(**Below**) A typical scene of night maintenance work on an F-16A.
(*Michael Hagerty*)

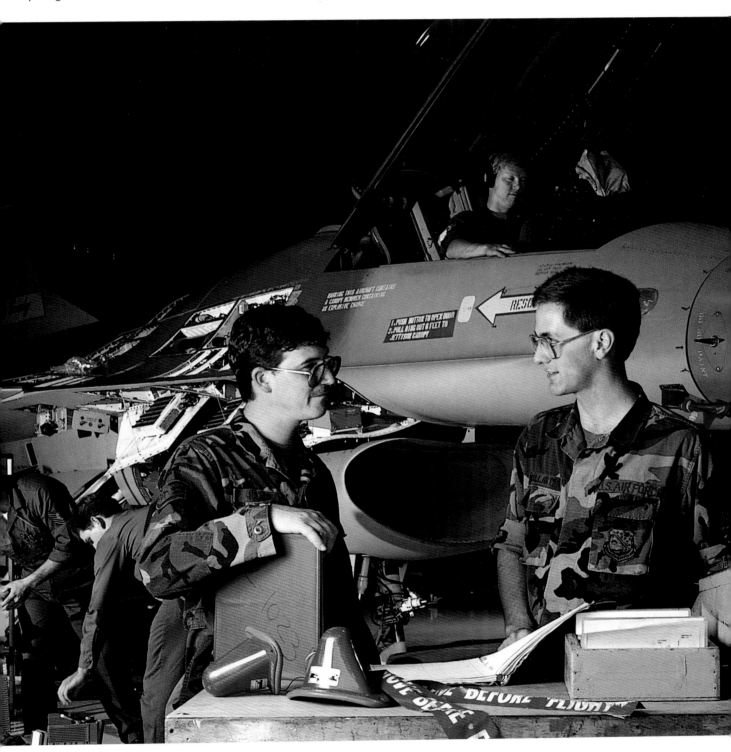

(Left top & bottom) Two angles on the front cockpit of a two-seat F-16D block 25 (83-1181) of the 184th Fighter Group, Kansas ANG; block 25s are the earliest C/D-models. Again, note the distinctive sidestick controller situated at the pilot's right hand. "It's hot," says Lt.Col.Michael P.Curphey, who has flown plenty of other fighters with conventional between-the-knees control sticks; "It's responsive, and it takes only very slight in-puts to fling the airplane around the sky." The ACES II ejection seat reclines at a 30 degree backwards cant in the F-16; this is widely advertised as reducing the G-forces on the pilot -

which is true; but in fact the seat was fitted at this angle because there was no other way to get it into the very slender forward fuselage. The HUD is situated at the pilot's eye level; just below it are toggles and other controls for the APG-66 radar. Conventional flight instruments such as air speed indicator, altimeter and vertical velocity indicator are mounted in the lower console between the pilot's knees. *(Jim McGuire)*

(Below) A look straight up the intake of F-16D (83-1181), showing various blade antenna and AOA sensors/probes described earlier. Since these photos were taken the Kansas ANG "Jayhawks" have converted to the Rockwell B-1B Lancer strategic bomber. *(Jim McGuire)*

(Left) An F-16C (83-1131) of the 33rd Tactical Fighter Sqn., 363rd TF Wing lifts off from Shaw AFB, South Carolina in 1991; note the AGM-65 Maverick air-to-ground missile on the outboard pylon. The 33rd "Falcons" and their wing-mates of the 17th TF Sqn."Hooters" saw plenty of action in Operation Desert Storm - and so did the Maverick, the original "fire and forget" guided weapon. Some 5,000 were expended during the Gulf War, typically against enemy armor, with a kill ratio of better than 90 per cent. The 363rd has since been redesignated the 20th Fighter Wing.
(David F.Brown)

(Bottom left) Armed with Sidewinders and Mavericks, an F-16C of the 302nd Fighter Sqn., 944th Fighter Group, Air Force Reserve (LR tail code) at Luke AFB, Arizona, is posed for a portrait with its most important feature: the pilot. The manufacturer of this great fighter, now Lockheed Martin, anticipates that the end of the century may find F-16s still in production in Turkey and South Korea - though perhaps not in the United States.
(Bob Shane)

(Below) A Lockheed Martin AAQ-12/13 LANTIRN (Low Altitude Navigation and Targeting Infra-Red for Night) pod mounted on an F-15C of the 310th Fighter Squadron. This system, with terrain-following radar, automatic target tracking, and laser rangefinding/designation, gives the pilot the capability to execute precision low-level attacks with weapons such as the Maverick under cover of darkness.
(Robert F.Dorr)

After a brief heyday during Operation Desert Storm, aircraft nose art has all but disappeared once more from USAF warplanes in the 1990s. "Tigger, King of the Air" is an exception, seen during a visit to Davis-Monthan AFB near Tucson, Arizona, for an exercise in January 1995. The artwork was applied with chalk, which endures surprisingly well during high speed flight but does not hold up in snow or rain. Since this F-16C block 25 Fighting Falcon (85-1416) belongs to the "Marksmen" of the 163rd Fighter Sqn., 122nd Fighter Wing, Indiana ANG and bears the FW tail code of Fort Wayne Municipal Airport, the cheery tiger probably did not long survive its return home. Note the golden tint of the RAM-treated canopy; radar-absorbent materials also line the air intake. (Robert F.Dorr)

CHAPTER 6
McDonnell Douglas F-15E Strike Eagle

(Top) The crew of the F-15E sit under a long, clear canopy atop a high, stalky undercarriage, well ahead of the wing leading edge; they enjoy superb visibility, both on the ground and in flight. The narrow track of the undercarriage (9ft, 2.78m) gives relatively little tolerance for crosswinds, however. Aircraft (88-1702) of the 4th Wing at Seymour Johnson AFB, North Carolina, illustrates the high "sit" and impressive size of the Strike Eagle - whose gross weight can reach 75,000lbs (34,019kg) with a full ordnance load for a long range strike mission. *(Robert F.Dorr)*

(Above) Flight refuelling is essential to just about every mission profile devised for the F-15E, as it is for most warplanes in the USAF inventory; endurance without topping up is usually about one and a half hours. Seen through the glare of another Strike Eagle's rear canopy, (88-1702) is taking on fuel from a Douglas KC-10 Extender near Seymour Johnson, at a rate of some 2,000lbs of JP4 each minute. Pilots find the Strike Eagle easy to handle "on the boom", although there are other times - especially in the airfield pattern - when the F-15E is not as forgiving as some other types. *(Robert F.Dorr)*

(Top) Strike Eagles are powered by two sets of engines: 25,000lb (11,340kg) Pratt & Whitney F100-PW-220s, or more powerful but less trouble-free F100-PW-229s. Engine failures in Alaska-based F-15Es led in 1994 to the grounding of many aircraft powered by -229s; the problem, tentatively blamed on stress-related cracking of the fourth stage turbine blades, was resolved after some months. A third model power plant, the General Electric F110-GE-129, is expected soon. Despite the intake cover, this view of (89-0488) shows how the engines are fed by straight two-dimensional external compression intakes. These embody the only use of variable geometry in the aircraft. Because the Strike Eagle may have to fight at high angles of attack, the intakes can "nod" up or down to keep the aperture facing directly into the airstream, maintaining a smooth flow of air to the engines. The intake angle can also be adjusted to prevent more air than necessary being taken in. The conformal fuel tanks bulging out behind the air intakes on each side of the fuselage are unusual in having the capacity to carry ordnance mounted on multiple racks. *(Robert F.Dorr)*

(Above) McDonnell Douglas F-15E Strike Eagle (88-1702) of the 334th Fighter Squadron, "Eagles", 4th Wing, in clean configuration for a simulated air-to-air combat mission from Seymour Johnson in June 1993. It is 63 feet long and 40 feet wide; the huge wing area and big dorsal speed brake mean no braking parachute is needed on landing. Note how the horizontal tail extends far beyond the exhaust rim, making it the rearmost section of the aircraft. *(Robert F.Dorr)*

Details of the wheel, and left and right wheel wells looking forward and upward. This particular ship, (90-0243), was assigned to Brig.Gen.Hugh Cameron, commander of the USAF 3rd Wing stationed at Elmendorf AFB, Alaska; this is the wing which would deploy to Korea in the event of a crisis. Photographed while visiting Hickam AFB, Hawaii, in September 1995, it has unique hubcaps - referring to the 3rd Wing's 90th Fighter Squadron - which were ordered by crew chief Staff Sergeant Frank Mallory, one of many Desert Storm veterans among the F15E's talented ground personnel. (Jonathan Chuck)

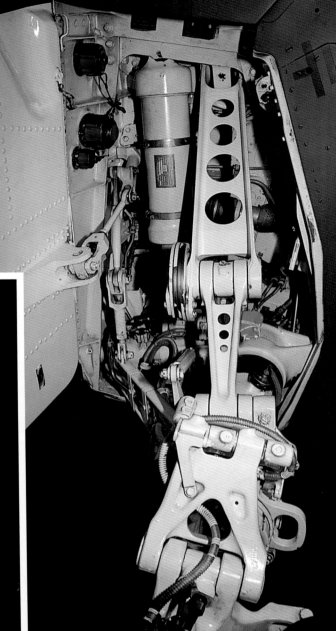

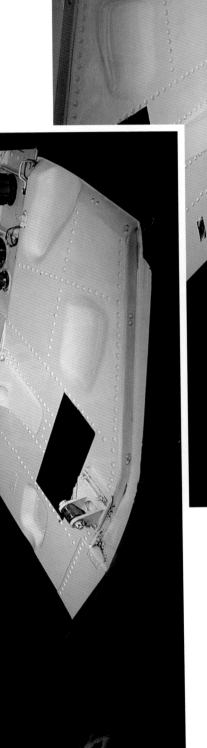

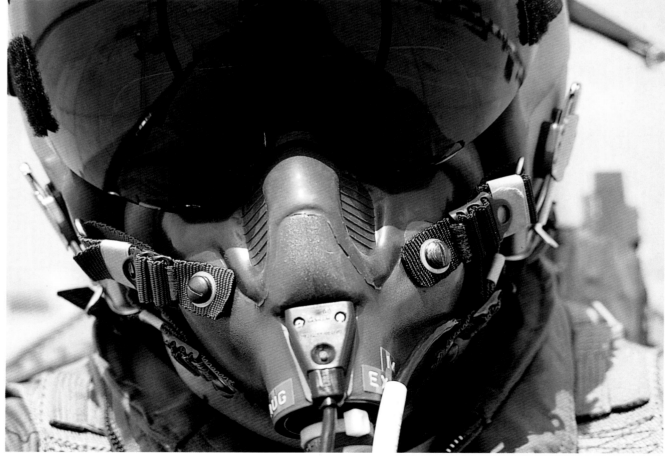

(Left top & bottom) Though the cockpits of most combat jets are cramped, and the author cannot claim to be quite as trim as he was when in the service, there is plenty of room in both the front and back seats of the F-15E. The pilots and weapons systems officers of the Strike Eagle - and the author, for this flight in the back seat in June 1993 - wear the HGU-55/P "Combat Edge" lightweight helmet with MBU-20/P oxygen mask, CWU-27/P Nomex flight suit and CSU-13/P anti-gravity pants. The parachute is integral to the McDonnell Douglas ACES II zero-speed, zero-altitude ejection seat. There are controllers in both crew positions with three ejection choices: "Norm" will eject both crew, whoever pulls the handles; "Solo" punches out only the man who initiates it; and "Aft Initiate" is self-explanatory. If you get a ride in the WSO's seat of an F-15E the pilot will tell you to select "Solo", just in case; he will prefer to make his own decisions about abandoning $44 million worth of aircraft.
(Robert F.Dorr)

(Above) It looks so simple - but the information available to the "wizzo" from his four multipurpose CRTs (and which he can route to the pilot's displays at will) makes a central contribution to the F-15E's superb combat performance. From his seat in back the WSO can manage the imagery and data of weapons status, target designations, monitoring of enemy tracking systems, flight control information, and communications. He operates the AN/APG-70 synthetic aperture radar and the Martin Marietta AAQ-13/14 LANTIRN infra-red navigation and weapons targetting pods; and even has his own minimal flight controls. The master panel at far right is the WSO's equivalent of the pilot's "up-front controls"; invisible here in a console left of his seat are a second button-studded hand control grip, and throttle levers.

The AN/APG-70 radar displays give the back-seater bird's-eye views of ground targets that are of higher resolution, and from farther away, than the images produced by previous systems. He can identify roads, bridges, and airfields from anything up to 100 miles (161km). As the Strike Eagle nears the target the image becomes sharper, enabling the crew to distinguish individual targets such as aircraft, tanks and trucks from perhaps 50 miles (80km). A feature of the radar is its ability to create and freeze high-resolution ground maps during quick sweeps of the target area lasting only seconds.

(Above and opposite)

General view of the upper and lower forward consoles (*Jonathan Chuck*).
The front office has "that new car smell", as one pilot puts it, and is testimony to the marvels of high-tech. The pilot flies in the HOTAS mode (Hands On Throttle And Stick), controlling the aircraft with fingers and thumbs while taking in key information from glowing green symbols created in the HUD (Head Up Display) screen atop the cockpit combing - at their most basic these will indicate aircraft heading, air speed, attitude, altitude, missile priority status including range to target, etc. The three CRTs (Cathode Ray Terminals) which dominate the console provide multi-purpose displays of navigation, weapons delivery and systems operations selected from a 16-choice menu; the pilot or WSO can punch up displays for air-to-air and air-to-ground radar, FLIR, HUD repeater, available weapons, etc., at will.

The central panel of "up-front controls" between the two upper monochrome screens comprises ten function buttons, six 20-character displays, four radio controls, and a 20-key data input section. Using this panel (which replaces many more of the individual instruments and controls which used to clutter the cockpits of the Vietnam era) the pilot can program all the subsystems that need information - the radios, INS, TACAN, ILS, IFF, etc.

(Above & over page) The color screen at low center typically displays e.g. the artificial horizon; or a "moving map" Tactical Situation Display coupled to the Inertial Navigation System, showing current position, course, targets, aiming points, etc. Left of this are analog dials giving basic flight information, and right of it an engine monitor display and fuel management controls. The console to the right of the seat houses, among other things, the integral starting system (JFS); that to the left, the throttles in their brush-edged slots, FLIR navigation controls, etc.

The HOTAS concept allows a pilot manoeuvering under high G-loads to control essential systems without ever taking his hands off the throttle and control column. The throttles incorporate buttons and switches for missile and gun selection and target designation, radio, radar, speed brake, laser rangefinder, chaff and flare dispenser operation, and air-to-air interrogation. The central stick grip incorporates controls for the individual CRTs and the data programmed for each, the trim, countermeasures, priority targetting and weapons management - including, of course, the gun button.

(Top) The arid mountains in the background will be a giveaway to many: this portrait was snapped at Nellis AFB, Nevada, close to the casinos of Las Vegas. F-15E (91-0604) of the 48th Fighter Wing, stationed at RAF Lakenheath, England (LN tail code), taxies into the "last chance" ordnance area for post-flight checks on the occasion of Air Combat Command's annual air-to-ground combat rehearsal, "Gunsmoke 95". This aircraft, one of four sent to the exercise by the 48th Wing, carries training rounds which will be "scored" to determine the bombing and shooting accuracy of the crew. Note the full fuel load; the blade antennas for radar warning receivers located beneath the forward fuselage; and the fact that even the crew's helmets are finished in low visibility gray as part of the overall effort to reduce the aircraft's vulnerability to heat-seeking missiles.
(David F.Brown)

(Left) See pages 118-119 for captions.
(Hans Halberstadt)

(Above) All Strike Eagle take-offs are performed on full afterburner, the twin 25,000lb-thrust engines each burning fuel at a rate of 40,000lbs an hour; the heavy F-15E lifts off at the head of a 30-foot flame at between 160 and 185 knots depending on weight - around 40,000lbs clean, nearly twice that with a full warload. The author can attest that take-off in an F-15E is a real kick in the pants; this is, after all, among the first generation of warplanes with a thrust-to-weight ratio so great that it can actually accelerate while flying straight upwards. Here (86-0190), taking off for the Gunsmoke 95 competition at Nellis, displays the black and yellow fin chequers of the 57th Wing.
(David F.Brown)

(90-0243) is the Strike Eagle assigned to the commander of the 3rd Wing at Elmendorf AFB, Alaska; Brig.Gen.Hugh Cameron is photographed during a visit to Hawaii in September 1995. His aircraft - parked with the huge dorsal speed brake open - displays the standard colorscheme for the F-15E, sometimes called "penetrator gray", in this case with the tail code and wing designation highlighted in white, and a discreet flash in the different squadron colors at the fin cap. The pilot's and WSO's names are painted below the cockpit on the left side in the "eagle" panel that has become standard in the F-15E community; the ground crew's on the right, in gray on solid black; and apparently, the wing's squadron emblems in black on the right fuselage behind the intake. In the side view, note the gun port at the root of the starboard wing for the General Electric (now, Lockheed Martin) M61A1 Vulcan 20mm rotary cannon, adopted after after earlier plans to develop a 25mm weapon for the Eagle failed to to reach fruition. The general's aircraft is carrying underwing ferry tanks.

In the head-on view, note the LANTIRN pods on the forward pylons. The navigation pod (starboard) includes terrain-following radar and wide field FLIR; the targetting pod (port) has a separate FLIR, laser rangefinder/designator, automatic target tracker, and automatic Maverick missile "hand-off". The indistinct mass right of the nosewheel is just a crewman's gear hung on the lightweight extending ladder; and note the "last four" of the serial number painted inside the nosewheel door. In the close-up, note details of the HUD screen (which appears green from the front), the huge canopy, and the upper part of the ACES II seat (the olive drab item on the console ahead of the WSO is his helmet bag, not part of the installation). The crew's high seating position relative to the canopy rails contributes to the superb visibility.

At the time of Cameron's visit to Hickam Field for the celebration of the 50th anniversary of the end of World War II, the US Senate was debating a US Air Force request for a small additional order of Strike Eagles which would keep the production line at St.Louis, Missouri alive after current orders for Israel and Saudi Arabia were completed.
(Jonathan Chuck)

(Above) The scale of the spectator in the bushhat gives a good impression of the menacing size of the F-15E in this close-up of the intricate tail feathers for the twin Pratt & Whitney F100-PW-220s. The elaborate exposed mechanism for expanding and contracting the size of the afterburner/exhaust nozzles to control thrust looks like a product designed by a mad professor for a 19th century iron works, but is much lighter than it looks and is exceedingly practical. The motors that drive all the control levers make a strange and very audible howling noise. F-15E (90-0229) of the 334th Fighter Sqn., October 1991. *(Jim McGuire)*

(Top right) From this angle, a "clean" F-15E (88-1702), carrying only an AIM-9M Sidewinder short range heat-seeking air-to-air missile training round, shows off its principal wing ordnance pylon, plus a myriad of shackles under the semi-blended fuselage for LANTIRN pods; the AIM-7M Sparrow long range radar-guided air-to-air missile; or its replacement, the AIM-120A AMRAAM long range "fire-and-forget" air-to-air missile. Note the disparate sizes of the radar warning receiver aerials atop the fins. This sortie from Seymour Johnson in June 1993 is flown by Major Bill "Hoppy" Hopmeier of the 334th Fighter Sqn.,"Eagles", part of the 4th Wing (since redesignated 4th Fighter Wing); the back seat is occupied by Lt. Col. Andy "Gecko" Gecolosky, a WSO with instructor/examiner skills. Seymour Johnson is

not far from Kill Devil Hill, where Orville Wright made history's first powered flight on 17 December 1903; note the white "Wright Flyer" emblem on the inside right fin - originally worn by KC-10 tankers at the base, and later picked up by the Strike Eagles. *(Robert F.Dorr)*

(Right) The talented camera of Dave Brown, American correspondent for Air Forces Monthly, brings us a typically sharp portrait of the 57th Wing's "boss bird", assigned to the USAF Weapons School. The WA tail code (probably inspired by the term "weapons acquisition") is highlighted in white, signifying an aircraft earmarked for somebody with seniority. The crew are buttoned up and ready to roll, with AAQ-13/14 LANTIRN pods on the pylons. At 63ft the Strike Eagle is nearly as long as the old Douglas DC-3/C-47/Dakota twin-prop transport of the 1940s, and has about the same empty weight; and since it is 18ft high, the Eagle's crew are about the same distance above the concrete. Apart from the affection inspired in their crews, however, that's about where all comparisons cease. *(David F.Brown)*

(Left) The right fin cap of an F-15E of the 550th Tactical Fighter Training Sqn. at Luke AFB, photographed in April 1989, is festooned with high-drag gadgets which were not part of the original design of the aircraft but which are essential to the mission. The spike-shaped radar warning receiver protrudes well forward of the leading edge. Behind and below are an electronic countermeasures (ECM) device, and a red navigation light. This portrait is a glimpse into history, since the squadron

no longer exists, and the base is no longer an F-15E operator: the Strike Eagle RTU (Replacement Training Unit) has since been moved from Luke to Seymour Johnson.
(Norris Graser)

(Bottom left) Attached to the conformal fuel tanks, the relative complexity of the F-15E weapons pylon is in part the result of its versatility: the Strike Eagle is "wired" to carry just about every item of ordnance in the USAF inventory. This is the standard Mk 82 500lb (227kg) general purpose

bomb, the most widely deployed gravity weapon in the armory. Latches within the rack assembly mate with fittings on the bombs; the ordnance is released in sequences programmed in by the WSO.
(Norris Graser)

(Below) Another view accentuating the differences between the radar warning receivers atop the twin fins. The colorful fin cap design of (87-0169) belonged in June 1990 to the 461st TFT Sqn.,"Deadly Jesters", then part of the 405th TFT Wing, the F-15E RTU at Luke AFB,

Arizona (LA tail code). Its identifying markings have disappeared with the massive reorganisation which shook the US Air Force in the 1990s. The Tactical Air Command badge survives, though the command does not; it is retained by Air Combat Command, which replaced TAC on 1 June 1992.
(Jim McGuire)

(Top) At time of writing the 333rd Fighter Sqn. were the newest operators of the F-15E, having "stood up" as the fourth squadron in the 4th Fighter Wing (formerly, 4th Wing) at Seymour Johnson near Goldsboro, North Carolina. Apparently fresh out of the paint shop, this "mud hen" (as the press sometimes dubs the Strike Eagle, from its low, dirty mission) was photographed at Nellis AFB in October 1995 during Air Warrior, an ongoing exercise hosted by the 549th Combat Training Sqn.; (89-0473) displays the red fin cap, unit designation, and discreet light gray tail highlighting which mark the squadron commander's "boss bird".
(David F.Brown)

(Above) A final image; and what more fitting than a leader and his wingman, photographed shortly after the Gulf War in 1991? The tail code and yellow fin cap identify respectively the 4th Fighter Wing out of Seymour Johnson, and the wing's 336th Fighter Sqn., "Rocketeers". As the 336th Tactical Fighter Sqn. this was the first Strike Eagle outfit to deploy to the Gulf, in August 1990. During the six weeks of the shooting war which began on 16 January 1991 the "Rocketeers" flew 1,088 combat missions totalling 3,274 combat hours, expending six million pounds of bombs and missiles on Iraqi bridges, batteries, armor, troop concentrations and missile sites. They lost one aircraft during training, and two shot down by ground fire; four aircrew were killed, and four captured. In an air-to-air victory which was never officially credited, an F-15E of their sister squadron, the 335th TF Sqn. "Chiefs", brought down an Iraqi helicopter on 14 February 1991 - with a laser-guided bomb....
(David F.Brown)